Exploring Political Ecology

This book explores some of the conditions and underlying causes of the multiple environmental crises facing humanity. Rooted in anthropology, but multidisciplinary in scope, it surveys the many socio-cultural and socio-economic errors, foibles, and follies that brought us to these circumstances. Crucially and uniquely, it outlines an array of viable and practical solutions, some of which are radically different from the current status quo and cultural expectations. The first chapter canvasses the emerging, interdisciplinary field of political ecology, then Part I examines details and trends in agriculture. Part II portrays the threats posed by carbon dependent and combustive technologies as well as the hydro and nuclear energy systems now powering the majority of human actions in developed parts of the world and expanding beyond. The third part turns to consider solutions, including green new deals, de-growth policies, localization, agroecology, alternative energy systems, and many more possibilities. The conclusions engage with urgent moral and legal issues and outline social movement strategies—all related to our collective neglect of climate change—and then finally speculate upon possible futures. This book is key reading for researchers and students interested in climate change across the social and physical sciences and humanities.

Alexander M. Ervin (PhD Illinois) is Professor Emeritus, Anthropology, University of Saskatchewan where he taught and researched for 51 years. His specialities include environmental anthropology, socio-cultural change, and applied anthropology. He is a past president of the Society for Applied Anthropology. His books include *Canadian Perspectives in Cultural Anthropology* (2001), *Applied Anthropology: Tools and Perspectives for Contemporary Practice* (2005) and *Cultural Transformations and Globalization: Theory, Development, and Social Change* (2015).

Exploring Political Ecology
Issues, Problems, and Solutions to the Climate Change Crisis

Alexander M. Ervin

NEW YORK AND LONDON

Designed cover image: Getty Images

First published 2025
by Routledge
605 Third Avenue, New York, NY 10158

and by Routledge
4 Park Square, Milton Park, Abingdon, Oxon, OX14 4RN

Routledge is an imprint of the Taylor & Francis Group, an informa business

© 2025 Alexander M. Ervin

The right of Alexander M. Ervin to be identified as author of this work has been asserted in accordance with sections 77 and 78 of the Copyright, Designs and Patents Act 1988.

All rights reserved. No part of this book may be reprinted or reproduced or utilised in any form or by any electronic, mechanical, or other means, now known or hereafter invented, including photocopying and recording, or in any information storage or retrieval system, without permission in writing from the publishers.

Trademark notice: Product or corporate names may be trademarks or registered trademarks, and are used only for identification and explanation without intent to infringe.

ISBN: 9781032801469 (hbk)
ISBN: 9781032799070 (pbk)
ISBN: 9781003495673 (ebk)

DOI: 10.4324/9781003495673

Typeset in Times New Roman
by Deanta Global Publishing Services, Chennai, India

Contents

Preface   viii
Acknowledgments   x

1   What Is Political Ecology?   1

PART I
Agriculture   11

2   Overview of Global Food Production: Pressures to Industrialize   13
High Modernism and Contemporary Agriculture 13
A Major Critique of Industrial Agriculture 16
The Green Revolution 20
*Political Motives for the Green Revolution and Other Global
    Agricultural Policies 22*
Value Chains 25
Foreign Land Grabs 26
Conclusions 27

3   Field Crops: Grains and Soy   32
Corn 32
NAFTA, Corn, Mexico, and the United States 39
Soy 43
Conclusions 47

4   Livestock Production   49
*The Virtual Extinction of the American Buffalo and the Rise of the
    Beef Industry 49*
Cattle Ranching and the Brazilian Amazon 55
Pork Production 57
Conclusions 60

5 Who Really "Feeds a Hungry World"?   63
Industrial Food Chains vs Peasant and Small-Farmer Food Webs 65
La Via Campesina and Food Sovereignty 68
Conclusions 72

PART II
Energy   77

6 Coal   79
Overview of the Significance of Coal and Its Environmental, Social, and Health Impacts 79
West Virginia and Mountain-Top Removal Coal Mining 83
Conclusions 89

7 The "Devil's Excrement"—Petroleum   91
Petrostates 92
The Alberta Tar-Sands 95
Conclusions 104

8 Hydroelectric and Irrigation Dams   108
Environmental, Health, and Social Impacts of Major Dam Construction 109
Conclusions 114

9 Uranium and Nuclear Power: The Case Against   117
Uranium and Nuclear Power in Canada 119
Uranium in Saskatchewan 120
The Problem of Nuclear Waste 130
The Search for a Nuclear "Repository" in Canada, Focusing on Saskatchewan 131
Conclusions 134

PART III
Solutions   139

10 Transforming our Political Economies and Dealing with the Issue of Constant Growth   141
Green Growth 142
Green New Deals 145

Degrowth 149
*Politically Who is Going to Take Responsibility for
    Solutions? 156*
Conclusions 159

11 Some Solutions in Agriculture   163
Agroecology 163
Regenerative or Restorative Agriculture 169
Natural Systems Agriculture and Perennial Grains 175
Some Other Considerations and Conclusions 180

12 Renewable Energy Solutions   182
Sources of Renewable Energy 185
Transportation 192
Buildings 194
Industry 195
Storing Renewable Energy 196
*Bringing It Altogether through Redesigning the Grid with Major
    Diversifications and Nested Localization 198*
Worrisome Caveats and Conclusions 200

PART IV
Conclusions   205

13 The Big Moral Question   207
"Buck-Passing" 207
*Morality, Law, and Some Issues Concerning the Young, Future
    Generations, and Nature 208*
Movements and Mass Action—Manifesting "Climate X" 215
Conclusion 221

14 What Might Happen Next?   222
A New Green New Deal? *222*
*How Could People Respond to the Scenario of Global Warming
    and the Collapse of a Global Civilization? 225*
Conclusions and Final Words 230

References   *233*
Index   *247*

Preface

As an anthropologist I have always been interested in the big picture—the past, present, and possible destiny of humanity. Yet as a senior citizen, I am extremely worried about the future we are leaving our descendants. I decided to document these concerns within the framework of political ecology—a synthesis principally drawing from the social sciences of anthropology, geography, sociology, and political economy but also from the humanities (here, history, and moral philosophy), law, and necessarily partly from the natural sciences including biology, agronomy, physics, atmospheric chemistry, toxicology, public health, engineering focused on energy and power production, and others where relevant.

The book is highly critical of existing political and economic systems as well as the current practices of industrial agriculture and energy extraction and use. I could have given them more of their fair due as trends of modernity and prosperity especially for those of us in the Global North. I did not because I believe as do many, while conceding that they have greatly benefited some sectors of humanity for a few centuries, that, unless halted, they are leading us all on paths to possible extinction or at least to the collapse of a global civilization accompanied by intense suffering. Besides my academic and activist interests, the writing of this book was motivated by what I consider has been done by us as crimes against the young, future generations, and people in the developing world. This will be elaborated more clearly in Chapter 13.

I originally came by my interest in environmental anthropology beginning in the mid-1960s through the works and teachings of Julian Steward and Demitri Shimkin at the University of Illinois (one of my two alma maters, the University of Toronto being the other) where they also had joint appointments with the Department of Geography. For fairly obvious reasons, my earliest research in the demanding Canadian and Alaskan arctic and subarctic regions reinforced for me the significance of ecology in human affairs. For almost 40 of my 51 years at the University of Saskatchewan, I taught environmental anthropology as a mixture of cultural and human ecology.

Then in the 1990s, while serving my first term on the Board of the Society for Applied Anthropology (SfAA) as editor of *Practicing Anthropology*, I became aware of the work of Barbara Rose Johnston (ed. 1994, 1997) under contract through the SfAA as internationally leading a team of political ecologists about environmental problems experienced by local communities under the onslaught of extractive intrusions by governments and corporations. Barbara's work and inclination toward social justice have served as my more direct inspiration and conduits to political ecology. I then concentrated my research, teaching, and local activist participation through such critical approaches. A course I taught specifically on political ecology from 2009 through 2022 serves as the framework for this book. Beyond academia, from 1998 through to the present, I have been engaged in environmental campaigns primarily involving agriculture and energy—the two main domains that I use here to illustrate the principles of political ecology and underscore humanity's predicament.

Political ecology rests upon an essential foundation constituted by the older discipline of political economy. In that regard I have been influenced by a wide variety of its practitioners including some self-identified Marxist ones who have made contributions to the field. But for me the most influential source has been E.F. Schumacher through his (1974) *Small Is Beautiful: A Study of Economics as If People Mattered* and as further exemplified by such inspiring chapters as "Buddhist Economics." This is especially true when it comes to suggestions about applying solutions locally.

Finally, as an applied anthropologist (Ervin 2005), I find myself obsessed with exploring policy solutions to the problems documented in this book—thus the extensive suggestions discussed in Part III and the conclusions. For far too long governments and businesses have relied on growth-enthusiastic neoclassical economists for advice. We need approaches that are radically different—*based on degrowth and the eventual establishment of localized relatively steady-state economies*. Besides the emergence of the field of ecological economics led by the late Herman Daily and Joshua Farley (2011), there are anthropologists, including Jason Hickel (2021) and Arturo Escobar (2015), involved in the advocacy and design of such systems. While far from being popular and now completely ignored by governments, they will probably be mandatory as policies for survival when the full impacts of climate change hit young people and future generations. As one of my married daughters, Samantha, recently put it, and as most young people realize, "the Canadian Dream is over." Social scientists have a role to play in preparing young people for such events and to help to design small regional, community-based adaptations in preparation for the coming upheavals of this century. At the core, we desperately need some alternate forms of political economy.

Acknowledgments

First, I am thankful to my friend and former colleague Clint Westman for his advice on Chapter 7 about Petroleum with the usual caveat about responsibility for potential errors or misinformation not being his. Clint's policy-oriented research (Westman and Joly, eds. 2019) on the tar-sands of Alberta in the context of consultation between First Nations communities and oil companies has been the major academic contribution for understanding these highly controversial and damaging projects.

The folks at Routledge deserve my thanks. Most of all, much gratitude goes to Meagan Simpson, PhD, acquisitions editor—both Routledge and anthropology are lucky to have her service. Her enthusiasm for this project along with her prompt, cheerful, and positive professional communication and initiative have been a balm for this author. Recognition and many thanks should be given to the productin team who through their very hard and detailed work have brought this book to print.

Next, it is an honor to acknowledge some of the primarily non-academic environmental activists that I have worked with over the last 25 years in a series of actions that you could say represent political ecology as practiced at the ground or social movement levels. In many ways, these people are exemplary in their selfless commitment to environmental issues in Saskatchewan and the Canadian Prairie provinces in general. They have all been highly inspirational and for me serve as a convivial network of like-minded. These include people in movements against genetically modified crops, pesticides, and livestock factory farms; the proposed Meridian Dam on the South Saskatchewan River; the Alberta Tar Sands; energy concerns through The Coalition for a Clean Green Saskatchewan, Green Energy Project—Saskatchewan, and the Committee for Future Generations all in our opposition to uranium mining and nuclear reactors; campaigns against uranium giant Cameco's Canadian and Saskatchewan tax avoidance strategies; concerns about my university's cozy relationship with Monsanto, the controversial agribusiness giant, as well as Cameco again, one of the world's largest uranium companies; and people within the New Green Alliance, Saskatchewan's original green party, in which I was a member (1998–2010) and twice ran as a quixotic candidate in provincial election campaigns;

along with projects as a board member of the Canadian Center for Policy Alternatives—Saskatchewan Branch.

Those named here, often in overlapping campaigns and in no particular order, represent among Saskatchewan's most progressively aware Settler, Métis, and First Nations people. So, in this regard, my thanks to Don Kossick, Lynn Oliphant, Rick Sawa, Darrin Qualman, Cathy Holtzlander, Gary Smith, Brenda Goldsworthy, Karen Weingeist, Dave Geary, Doug Blackport, Peter Prebble, Marius Paul, Candyce Paul, Max Morin, Debbie Mihalicz, Jim Penna, Emily Eaton, Erich Keiser, Mary-Jean Hande, Peter Garden, Simon Enoch, Jan Norris, Neil Sinclair, Priscilla Settee, the late Dave Greenfield, Ivan Olynek, D'Arcy Hande, Mark Bigland-Pritchard, Jim Harding, Howard Woodhouse, Claire Card, Kathleen Solose, Tim Quigley, Richard Julien, and the late Len Findlay. Also, three friends and fellow anthropologists in the persons of Kendall Thu, Elizabeth Guillette, and Barbara Rose Johnston from the United States have joined us to assist on campaigns regarding factory farms, pesticides, and uranium mining, respectively.

Also essential to the writing of this book has been the support of Betty-Anne and the late Gordon Fraser on Big Island of Merigomish, Nova Scotia, who kept this citified, mechanically challenged, klutzy author logistically afloat while writing much of this book. As next-door neighbors at this beautiful but relatively isolated island along with them and my son JD, we experienced the devastation of Hurricane Fiona in September 2022 and were forced to adapt without running water and electricity for 12 days in its aftermath. That experience led to some speculation about the kinds of futures many will face as a result of climate change and that figures in the final chapter which suggests a kind of hunkered-down localized frugality that may be portended by the example of that event. Having no siblings myself, for over 35 years Gordon and Betty-Anne have served as dedicated uncle and aunt to my four children, To them—Samantha, John Donald, Maggie, and Jennifer—many thanks to you for your loving support to this geezer in his twilight years.

1 What Is Political Ecology?

We are both the asteroid and the dinosaurs.

Anon

Political Ecology is concerned with environmental injustices and inequalities.

Alf Hornborg 2015: 378

We have entered an era that some have labeled the Anthropocene (Lewis and Maslin 2015; Kolbert 2014). In it, the dominating, even geologically equivalent, forces upon the earth and its biosphere have been the result of human action. There is a paradoxical aspect to it all. This is because this human capacity that, in the minds of some (Latour 2017), can now actually be equivalent to volcanoes, earthquakes, and even tectonic plate action is also a form of destructive agency. We are becoming aware that it has the possibility of leading to our own extinction, thus metaphorically we are the "asteroid" and the "dinosaur" simultaneously. Although collectively and historically true, it needs to be repeatedly underscored that not each and every human being is, by any means, equally complicit. Some are very much responsible and some are not at all serving much more as victims than as perpetrators. Among the rest of us, there is still plenty of responsibility that can be distributed in varying degrees of culpability and victimhood, often with nuanced mixtures of both.

These issues get quite complex when we examine them in light of social and political factors and those of wealth and power. Consider the Marshall Islanders: after already suffering from U.S. nuclear bomb testing, they are now subject to the likely disappearance of their homelands due to flooding from rising sea levels resulting from global warming (https://www.nationalgeographic.com/science/article/marshall-islands-climate-change-floods-waves-environment-science-spd). Compare them to the infamous Koch brothers or successive governments in Alberta, Canada. The former (Mayer 2017), beyond the fact that their companies have been frequently fined for major petrochemical pollution, are responsible for massively funded, climate denying, think tanks, and the latter (Nikiforuk 2008; Urquhart 2018) for promoting and extracting the dirtiest carbon-producing petroleum—bitumen from tar sands—on the planet that negatively influences the conditions the Marshallese now face. To add to all

DOI: 10.4324/9781003495673-1

of this, how about the complicity of the average citizen, including myself, in Saskatchewan where I live, which has a per capita CO_2 release into the atmosphere of almost 70 metric tonnes a year—among the very highest in the world (https://www.conferenceboard.ca/hcp/ghg-emissions-aspx/)? Although the average Saskatchewanian might be at least partly absolved by the fact that much of that excess is due to the heavy concentration of extractive industries, such as petroleum, in the province—but only partly so.

So, there is now a great deal of conflict and perplexity about all of this and, in understanding its importance and complexity, there is confusion about older concepts and frameworks that are outdated or may no longer apply. Can we really justify separating culture from nature? What might be the point of arbitrarily dividing the understandings of this new regime into very distinct specialties of study separating the human from the non-human? Can we continue to justify this, when so many of the transformations arise first from various human ways of thinking and are directed through actions perpetuated by particular human institutions and technologies—in other words, what we have called culture? Shouldn't we at least try to bring it all together in one framework? So, to these ends, political ecology as an interdisciplinary subject has emerged since the mid-1980s when our awareness of these worrisome issues began to become more widespread in both scientific and public consciousness.

Alf Hornborg (2015) captures one of the key features of our subject—human environmental justice—*advocacy and critical commentary* on a grand scale and it tends to be activist rather than merely academic. Action, even if only in the form of vigorous and rigorous advocacy, is called for more than mere contemplation and the pretense of perfectly neutral "objectivity." The subject documents past and current physical transformations of the global biosphere by humans through their technology, social institutions, and cultural belief systems. It focuses then on environmental crises and the social injustices associated with them.

The subject is clearly *political* in its treatment of ecology. It is unavoidably rooted in conflict and identifies adversaries and courses of action that need to be taken. It politicizes, as we shall see many times in this book, issues that, through the manufacturing of consent, have been generally ignored, brushed over, or perceived as leaving no choice but to continue in customary ways that seemed natural. For instance, is it true that we are in a continuous war with nature and must therefore use our best weapons such as chemical pesticides in order to feed ourselves? Is it true that we in the developed world, as seems to be our right, have grown so accustomed to certain minimal levels of energy consumption, and that there is no choice but to maintain them while expanding them to account for population growth and those in the developing world aspiring to join us in comfort and affluence? Is it true, given the "self-interested nature" of human beings, that the best way of managing resources is through private property ownership, and the opposite alternative could only be totalitarian communism?

Academically political ecology is eclectic, still a work in progress, and a sprawling one that spans and tries to integrate a number of disciplines. It does not have a single dominating theory, a declared range of methods, and does not belong to any one traditional subject. Political ecologists deal with vital environmental issues mainly through the combined perspectives of the social sciences, but also the humanities and, of course, the natural sciences. Anthropology, human geography, and allied subjects such as history are especially important. It combines insights from these subjects with observations from biological ecology, climatology, physical geography, the health sciences, agronomy, many relevant orientations from the natural sciences, and the humanities (philosophy and history primarily). So, the subject is integrative and holistic, which is a major and essential contribution when the vast majority of subjects restrict their boundaries of enquiry and tend to use reductive and linear methods that mask the complexities of cause.

Political ecologists take the position that the majority of current environmental crises such as global warming, mass extinctions, and land and water degradations are caused by human actions. These have been historically constructed through powerful social institutions, historically through the influences of capitalism, industrialization, the agendas of nation-states, and by the push for globalization in economic transactions. In doing all of this, it shows that the impacts of these transformations are almost always pointedly unequal with some peoples paying much more of a damaging price in contrast with others who clearly benefit the most. At the very forefront of these causes are political and economic ones, although other institutions and ideologies may also contribute.

The economic faith in growth of a nation's gross domestic product through mass consumerism; the value placed on the "new"—the ideology of modernity; or the presumed desirability of increasing personal and collective wealth; or the notion of the "market" as a primary value and the director of human decisions; or the Western Judeo-Christian notion of humans having dominion over nature; or the widespread presence of patriarchy as justifying the conquest of Mother Earth (Merchant 1980), all can be examples of culturally based ideologies that may lead humans to exploit the material resources of the Earth's environments beyond renewability and toward damaging environmental risks.

Yet more than ideologies, the earliest emphases taken in political ecology have been on the actions themselves and the institutions and material circumstances through which they have been accomplished—in these cases largely *political* and *economic* where the most significant dimension of *power* tends to be most evident. It is for this reason that the field of political ecology is essentially influenced by the much older subject known as *political economy* that is also critical in nature. In part, political ecology could be regarded as a merger of political economy with previous attempts to account for human environmental actions within social and cultural frameworks. Yet there are significant differences in that the newer emergent paradigm has a much broader scope and is not restricted to localized situations of near equilibrium as with horticulturalists, foragers, peasants, and pastoralists; almost always has an overriding critical tone to any analysis; and deals with the consequences of industrialized primary industries such as agriculture.

For more information on the emergence and underpinnings of political ecology from an anthropological perspective, I recommend the article titled "Political Ecology—Introductory Essay" (Greenberg and Park 1994) in the inaugural issue of the opensource online *Journal of Political Ecology*. As well there are other important references from geography that recount the beginning, scope, and agenda of our subject (see, for instance, Watts 2015; Bryant 1992; Blaikie and Brookfield 1987). It should be also noted with regard to the subject's stimulus in association with political economy that the latter was re-invigorated through works associated with the notion of a globally integrated World Economic System (see Frank 1975; Wallerstein 2004; Wolf 1982), a major transformation that began in the 1500s through European imperial powers and mercantilists rapidly establishing trade routes, exploitative extraction systems, and continental exchanges of commodities of furs, spices, fish, precious metals, tea, coffee, sugar, timber, so forth, and most shamefully of human beings as slaves. This approach to political economy extends the analysis through the emergence of industrial capitalism and contemporary globalization. It is notable especially with Eric Wolf's (1982) work for showing the complexities of social and cultural transformations occurring as all peoples become interconnected in intricate systems of differential power and influence and constantly subject to dynamic changes forged mainly by economic and political forces. Still missing, though, was a solid exposition of how this World Economic System was changing environments and peoples' relationships to them. That led to political ecology itself emerging in the 1980s—at first through the works of pioneering geographers such as Michael Watts (1983), Piers Blaikie (1985), and Suzanne Hecht (1985). Some later significant anthropological works were added by Barbara Rose Johnston (ed. 1994, 1997), Emilio Escobar (1999), and Alf Hornborg (2001, 2011, 2016). Our interdisciplinary subject continues to expand with several journals such as *Capitalism, Nature, Socialism*, a textbook on its geographical dimensions (Robbins, P. 2019), and several comprehensive handbooks (Perreault et al. eds. 2015; Bryant 2015).

Power differences—military, political, administrative, legal, financial, ideological, class, race, ethnicity, and others—often overlooked elsewhere in social science enquiries have always been the focus of political economy. Considered also are the formal institutions of governing societies—kingdoms, empires, nation-states, and their regional equivalents of provinces and states sometimes along with indigenous nations living in frontier regions. Governance and management of resources, for example, through legal institutions favoring private property or through customary but regulated sharing as in the *commons* are highly significant. Economic factors and institutions are always of high importance—technologies, markets, globalized trade, trade agreements, debt and credit, the role of banks and investment and financial firms, development, extractive industries, commodities and commoditization, factors of production, consumption patterns, corporations, and transnational corporations as well as the networked relations of particular people generating the actions and conditions that we will be discussing in this book. How these factors are all organized and practiced can have enormous influences in shaping environments locally and now globally as well as the status of the differentiated peoples dwelling within them.

In its multi-disciplinarity, political economy's offspring, political ecology frequently takes into account legal and policy factors. If, for instance, a nation's law denies subsurface rights to landowners that could lead to farmers and local residents being victimized as is the case in Saskatchewan where I live. Carter and Eaton (2010) report that the heavy concentration of shale fracking for oil on farmers' and ranchers' lands has led to fugitive emissions from gas flaring that contribute a large portion of the province's already high greenhouse gas releases. Also, each well, and the number is in the 1000s, requires on average 750,000 gallons of local fresh water that is permanently polluted with toxic chemicals and lost to the water cycle, and much remaining native prairie grassland is degraded. Farmers can do nothing to avoid all of this, payments for granting access to their land are minimal, and it is extremely hard to arrange for any clean-up after each well's operations are over.

Similarly, anthropologist Jason Hickel (2017) reports how international trade policies, such as the North America Free Trade Agreement, allow foreign corporations to sue national governments when their environmental laws protecting the health of communities threaten their "expected future profits." He describes several egregious cases of just that—with impoverished developing countries being forced to pay transnational companies millions of dollars. Governments are thus often reluctant to upgrade their environmental laws even when the need is obvious.

It is the case that these two realms, the political and economic, are deeply intertwined with networks of people constantly circulating, interacting, and subject to each other's spheres of influence, agendas, and agency both in competition and cooperation. Power here means being able to bring about action on the part of others. For example, there are many instances when manufactured products such as pesticides or food additives are quickly rubber-stamped as being safe by officials in government agencies because of vested interests and the mutual personal favors of people circulating in and out of government and private corporations associated with the agencies' regulatory mandates (see Vallianatos 2014). As well, there are institutions that act in semi-governmental ways by representing the collective interests of a large set of nation-states yet operate primarily in collusion in accommodating the interests of transnational financial and corporate interests. Some examples of these include the World Bank, the International Monetary Fund, and the World Trade Organization.

Classical political economy was known for its more holistic analysis of society. While it did place emphasis on the material economic and political realities of any society's social and cultural existence, its practitioners were well aware that the rest of the institutional clusters found in any society were shaped by and reinforced by these assumed primary drivers—but they, in turn, did have some feedback potential for shaping political and economic institutions. They could include the important realms of religion, education, communications and media, the arts, philosophy, scientific ideologies, and just about anything of which one could think. As we have become more familiar with our relationships with the nonhuman and non-social environment, we realize that these linkages are multiple and therefore many variables need to be taken into account when considering human interactions with their biophysical environments. Political economy with its more holistic approaches than its highly specialized offspring disciplines of political science and economics is a good start, but even then, the linkages need to be extended and that is where anthropology and human geography are useful.

Studies of development and social impact assessment that began in conjunction with environmental impact assessment have played a big role in setting strategies for political ecology as well as providing data for this field. During the 1960s, there was an emergence of a public consciousness about some extremely negative consequences of modernity and the developmental and industrial approaches to primary industries such as agriculture and fishing, or major projects such as hydroelectric and irrigation dams. These could include, for instance, the poisonous results of agricultural pesticide use on the survival of wildlife as well as on human health (Carson 1962). It could be illustrated by the destruction of large areas of water-fowl breeding from flooding wetlands for dam-serving reservoirs. This led to the legal requirements for studies to measure and describe impacts upon particular species and according to certain variables as well as non-living parts of specific environments that could be impacted. This soon resulted in similar requirements with regard to any human populations in the areas.

Many of these projects, such as dams, power plants, and oil and gas developments, were being proposed and built in rural hinterlands where peoples had their livelihoods based on long-term, traditional, or semi-traditional relations of foraging dependence on particular plants and animals. To what extent would these dependencies be disrupted? A multitude of variables—subsistence, employment, settlement patterns, spiritual relationships, and a host of other factors—had to be assessed. There was also a common-sense assumption that some populations would pay more of a price than others. Costs and benefits should be measured or described, and two choices tended to be available—first, if the costs were too high, then the project was supposed to be canceled, or second, there would be attempts to compensate in some appropriate way the populations that were subject to the higher costs. Unfortunately, many projects that should have been canceled were not. The harsh realities of vested economic interest and power often dominated, and there was neither cancellation nor appropriate compensation for damage. In reality, for much of the development that has been occurring in the world, neither environmental nor social impact studies were done. If anything, and most of the time, the subordinate populations have been more often seen as even "standing in the way of development." The prices paid have been enormous and as we shall see the critical studies of development or actually in many cases "underdevelopment" with considerable damage significantly motivated political ecologists.

Public advocacy based on scientific expertise tends to be frowned upon in academia. The standard bias enculturated through disciplinary training is that scientists, natural or social, should avoid popularizing or taking public positions advocating one side or another of a position. Instead, they are expected to provide value free, objective, carefully collected data that is relevant to an issue. They are expected to stay out of the way, while policy-makers with the authority responsible for the domain in question supposedly rationally make appropriate decisions based on the objective data. This tendency is rare to absent with political ecologists—they take strong critical positions of advocacy based on value standards championing of those who suffer the most from various forms of displacement, pollution, ill health, loss of livelihood, and many other circumstances. Political ecologists tend to be *political* in another sense of the word. Neutrality or its false facsimiles are rarely present in political ecology research. That being said, though, the standards of fieldwork data collection, analysis, and conceptual framing remain high within political ecology. It is by no means an exercise in making up fake information to justify a political position. The data has integrity as do the researchers generating it.

Social and environmental movements play a very large role in the emerging field of political ecology—in fact, political ecologists may well be supporters or even participants in such movements. Paul Hawken (2007) in his *Blessed Unrest* writes of what he calculates to be over a million large- and small-scale, regional, social, human justice and environmental movements. According to him, they metaphorically serve as the Earth's immunological system in their collective attempts to try to stem the excesses of environmental damage and provide healthy alternatives to the future. This relates to the controversial but appealing position of James Lovelock (1979) that of Gaia—that the Earth as a whole constitutes a coherent and whole self-sustaining organic entity.

The earliest more secular social movements began in the 1700s in the context of the nation-state and modernity. They took the form of political parties, anti-slavery movements, nationalist, social justice, and education, labor, and feminist movements. Generally, these were intended to bring about social and economic justice in previously class-based and authoritarian societies now undergoing enormous changes through industrialization and urbanization. Social movements articulating and making their claims upon the state and society have been the major sources of social change, especially in the 19th century, with many continuing in the 20th century.

In the 1960s, movements also started to trend toward matters of lifestyle rather than just largely economic or political claims (Touraine 1985). These newer movements included women's, gay liberation, and indigenous peoples. They also displayed a large upsurge in environmental protest and the seeking of alternative lifestyles. This is continued very much to the present with a new twist—that of the emergence of movements in the developing world among more traditional indigenous peoples who are attempting to preserve their cultures as well as the traditional balances in relationships to their environments. Both are under enormous threats because of large-scale intrusions and extraction projects in their territories. Although following some of the tactics of the usual styles coming out of Europe and Euro-American contexts, they are quite innovative in the culturally distinct ways that they make their claims.

Overall, the movements include large global interest groups such as Greenpeace, Rainforest Action Network, and 350.org that relate to a multitude of issues in many regions of the world or to issues overlapping in many regions such as global climate change that has been the purview of 350.org. Nationally, groups such as the Sierra Club in both the United States and Canada separately deal with sets of common issues such as global warming, deforestation, pollution and degradation, and so forth. But there are many thousands of locally distinct, bottom-up movements that are making a difference and need documentation in the context of resistance and seeking sustainability.

Political ecology has plenty of applied and policy potentials but it has yet to make significant inroads there. This may be due to its highly critical stance. While analyzing the dynamics of environmental crisis with methodological integrity, its observations are not likely to please those in power who make the decisions that have led to the predicaments we study. As a geographer McCusker (2015), who worked briefly with a major American government institution—USAID—that oversees much of American aid projects in international development, noted that the employees there, dominated by the field of economics, tend to be deeply infused with neo-liberal economic ideologies that perceive constant economic growth as the only relevant outcome. Working with social movements or progressive non-government organizations, for instance, that promote organic and other forms of agriculture in the developing world could be one possibility to make a breakthrough. The collaborations that social scientists have had while participating in the agroecology movement currently in Latin America where agronomists and peasant farmers work together and along with them to develop agricultural systems based on traditional small-scale systems in more harmony with local ecosystems is another one (Gonzalez de Molina et al. 2020). These innovative systems have deep intrinsic and social connotations and are alternatives to what are presumed to be the destructive consequences of large-scale, agro-corporation dominated, industrial agriculture. Yet for the time being, other than advocacy and working with social movements in common cause, the applied dimensions of political ecology are still more limited than is desirable.

So, in summary, political ecology is a critically aware, advocacy and social justice prone interdisciplinary project relying on the foundational dimensions of political economy. The main effort is to expose the human institutional dimensions that generate environmental crises. The field is highly holistic, similar in essence to anthropology and human geography, its two main participating fields. It can draw upon biological ecology, the health sciences, agronomy, engineering, policy science, law, sociology, geography, anthropology, history, and even other disciplines from the humanities now that their post-structural, constructionist approaches have had some lasting influences on a range of social sciences. For its research, it relies upon time-honored fieldwork methodologies that are anthropology's strengths in showing the unequal damages done to subordinated peoples by more dominant and powerful institutions and groupings. It focuses on costs and benefits. The costs can take the form of lost autonomies and sovereignty, lost or damaged ways of making a living, ill health and poor nutrition, poverty, and lower standards of living—overall generating dependencies and crippling underdevelopment. So, while taking into account significant amounts of biophysical information and analysis, political ecology draws heavily upon history and the social sciences to examine and explain the current state of humanity's relationships to its physical environments.

The book is divided into three parts. Part I will look at the vital realms of feeding ourselves through plant and animal agriculture. Here well over 95% of our food is derived through the soil and other gifts of Nature but is ever-increasingly industrialized, chemically dominated, and now being manipulated through genetic engineering. In its industrial form it is responsible for up to 30% of climate changing greenhouse gases. Agriculture is an important topic for our critique because it has been the source of the most extensive and consequential transformations of our planet and the most destructive of ecosystems serving other life forms.

Then in Part II we shift to the hugely environmentally threatening, largely carbon dependent and combustive but also hydro and nuclear energy systems now powering the majority of human actions in developed parts of the world and expanding beyond. The majority of climate change since the Industrial Revolution through atmospheric interference has been primarily due to the production and consumption of hydrocarbon fuels. These chapters detail human-caused environmental damages most especially through carbon releases and impacts such as "wastelanding" or the making of sacrifice zones along with the continuing effects of injustice for local victims. Together the consequences of the combined practices of industrial agriculture and energy production and use have been the main triggers for global climate change, pollution, and losses of biodiversity. As these trends continue, they severely limit options for future generations to a significant but yet unclear extent.

Potential solutions to these issues involving technical, behavioral, institutional, and economic ingredients will be the theme of Part III. My approach here will not be completely integrative—in the sense of presenting a grand theory with interrelated components for solving our multiple and very serious environmental problems. However, I will highlight certain over-arching themes—degrowth, reclaiming the commons, and much more attention to localization rather than globalization. It is because of the last of these factors—localization—that I have taken an "assemblage" approach in that citizens and policymakers would ideally select from the multitude of solutions that already exist to fit the realities of their particular bioregions and socio-cultural traditions rather than continuing with the hegemonic outcomes of imperialism and capitalist modernization. Finally, in the two concluding chapters, I feel compelled to explore some of the possible consequences facing future generations and the kinds of adjustments that might become necessary for sustainability. Surely, most if not all readers here can agree that there is an overwhelming moral imperative to act to significantly improve the circumstances of our descendants and today's young people.

Part I
Agriculture

2 Overview of Global Food Production
Pressures to Industrialize

High Modernism and Contemporary Agriculture

I begin this chapter with observations involving my home context, Saskatchewan a Canadian province founded in 1905 based on extensive grain agriculture and containing 43% of Canada's farmland. Back in the 1970s, I was conducting a research project in a Saskatchewan farming region noted for its high-quality soils and the successful production of canola, a lucrative oil seed, and its traditional crops of wheat and barley, and, at the time, fairly resilient mixed farms of grains and hog raising. Seeking to get a fuller sense of the complete local social organization associated with farming, I went beyond interviewing just farmers but also other occupations such as agricultural extension agents, pedigreed seed growers, grain elevator operators, implement dealers, and others. Just outside one of the larger towns was a federal government experimental farm. It had been notable for testing and developing quick maturing and quick drying "bearded" wheat and barley with high protein counts as well as some of the successful strains of canola.

I had a long and fruitful interview with two PhD crop scientists who were the principal researchers at the farm. In my naivety, I asked them if they had any interaction with farmers in the region and that is when one of them said, "I don't do any work for the benefit of any local farmers. I do my job for the consumer but mainly for agricultural corporations." They said they had no contact with farmers and that nothing of what they did was directed toward any economic benefit for them and their communities. They worked toward developing high yielding crops that could be used commercially and to give Canada marketing advantages internationally.

I found that a troubling revelation since I had perhaps had an overly idealistic notion of agencies serving rural areas working in holistic combination to maintain them as communities as well as furthering farming itself. As fieldwork proceeded, I became aware of local residents' senses of abandonment by the larger urban-dominated society and governments. They complained of the school consolidations beginning with the abandonment of the one-room school that had anchored rural "social districts" to those that were found in small towns being combined to larger towns requiring their children to be bused as much as

DOI: 10.4324/9781003495673-3

fifty miles away each day. Then there were the serious closures of grain elevators again into larger regional "inland terminals" run by large corporations such as the transnational giant Cargill rather than the previous farmer cooperatives and the loss of rail lines that served them. A bigger blow later came with the elimination of over four hundred regional rural hospitals under a consolidation policy of establishing twelve hospitals serving a heavily rural population in a province roughly the size of Texas. All of these actions left rural residents bitter and abandoned, along with the fact that younger generations were leaving the region with the typical local high school graduating grade 12 classes ending up elsewhere. The average age of farmers in the region tended to be in the late 50s and younger farmers, few as they were, had to rely on seasonal off-farm work and were constricted by huge debt loads brought up by the increasing cost-squeeze pressures of farming.

What has happened in Saskatchewan agriculture and the landscape of its community viability is typical of what is found in all of North America. In 1936 after a period of approximately 30 years of homesteading and establishing the system, there were 142,000 census farms in 1936. This was at a time also when there were slightly more horses in the province than there were people indicating a still high dependence on animal traction in farming before World War II when that changed quickly toward mechanization. By 2021 the number of farms had been reduced to 34,128 (https://www.farms.com/ag-industry-news/saskatchewan-in-the-2021-census-of-agriculture-836.aspx).

Although anecdotal, my son, who was supervising construction work in South-eastern Saskatchewan, negotiated with local farmers for field stone necessary for the walling of an underground storage for the coal ash project he was building, tells an interesting tale. He encountered two brothers who ran a gigantic grain operation of over 75,000 acres mainly with the assistance of specially imported laborers from Germany with whom they felt they could only trust with their multi-million-dollar investment of tractors, combines, and other machinery. Presumably a portion of the land was not owned directly by the brothers but rented from absentee owners and the land would not be contiguous but would be scattered about the countryside. On the home quarter owned by the brothers, a full 160 acres or one quarter of a square mile was devoted to the housing of metallic grain bins that are used for storing grain until prices would be advantageous to selling. This seems like the sort of scenario that one could imagine for the near future of grain farming and has been experimented at this scale before. It is hard to imagine that a farmer managing a farm totaling 120 square miles would be paying much attention to the long-term sustainability and environmental aspects of such an operation or have much time to devote to the local community.

Related to all of this, James C. Scott (1998) suggests that the ideologies and practices of "high modernity" have shaped much of what has driven the 20th century and are still huge in shaping our current millennium. Much of that is based on the successes of factory production and industrial development in the late nineteenth and early 20th century. In his book *Seeing Like a State*, a new stage began led by the emergence of agricultural engineering, a new discipline, and the enthusiasm of top-down planners, especially in the U.S. Department of Agriculture.

It was based on a high optimism that farms could be run similar to factories. The "Taylorism" principles (time-motion studies) placed faith that the work processes on a farm as on a factory could be run on simple procedures that involved the working of machines and that unskilled workers could learn those procedures. Agricultural engineers looked for particulars of a farm operation that could be standardized—the layout of farm buildings, tools and machinery, and any significant way that could be used that encouraged mechanization in the production of grain crops. Early on they were obsessed with increasing the scale of farms with the intention of mass-producing highly standardized grain commodities while at the same time searching for ways to reduce the per unit costs of producing food. At first, it was attempted indiscriminately by ever-expanding the size of operations in imitation of industrial manufacturing. Limitations were eventually discovered that led to a more continued emphasis on family-run operations. One early attempt in 1918 was the massive Campbell wheat farm in Montana with an operation of 95,000 acres that was established with public investment and subsidies from the U.S. government. The thinking of its founder was that farming was 90% engineering and 10% agriculture. Two hardy crops, flax and wheat, were grown. The equipment consisted of 33 tractors, 40 binders, and 10 threshing machines. Its approach had to be abandoned, but many of its elements were preserved and diffused to existing farms that drastically changed over time, and today the small family farm's preservation is once again in dispute. Yet as exemplified above in my son's anecdote, very large-scale farming is on its way back a hundred years later.

The modern agricultural realm is imagined and realized in a context where intense, market competitive pressures cause agronomists to develop crops that tolerate crowding—pack as many plants of a single species and variety of uniform size within a field as you can. This leads to the intensification of fertilizer use, most especially nitrogen and for plants that quickly respond and make use of those fertilizers. Maize or corn growing in the United States aptly displays these characteristics. Accompanying this has been the development of large supermarkets and very long supply chains that are now global. There are standardized routines of shipping and display that also lead to uniformities in particular marketing standards of shape, color, and texture.

There has been an emphasis on a small number of plant varieties out of the many thousands possible—for example, potatoes down to about five or six. Also, with the heavy emphasis on mechanization, farmers sought and were provided with such varieties that were compatible with their machinery. *So, significant diversity in plant regimes is contrary to modernization in agriculture and that does not bode well for long-term environmental adaptations.* With commercial hybrid plant breeding beginning in the 1920s, crops were designed to be harvested by machines with uniformity of size, shape, and ease of separating fruit or seed from the whole plant. The "supermarket tomato" bred in the 1940s is a good example—tough-skinned, low growing, and capable of being harvested by machines while still green and artificially ripened by ethylene gas while in route to the stores where they were sold four-to-a-package as winter tomatoes. Taste and nutrition were secondary in such cases.

A basic lesson that was learned through the Irish Potato Famine of the 19th century was the danger of crop disease when monocrop systems were prevalent. All of the plants in the field will have the same degree of resistance or lack of to a plant disease, and thus eventually disease could wipe out entire crops causing famines. Previously, the best way of dealing with this was to have multiple varieties such as traditional landraces of the same species or to have polycultural systems of many species. For biological adaptability and resilience, diversity is the enemy of epidemics. A means of adaptation in monocrop systems through agricultural science is the periodic development of hybrid crop varieties to maintain vigor. Yet even then the varieties have to be kept changing every five years or so as pests themselves quickly adapt to particular hybrids. Furthermore, accompanying the development of hybrids has been the development of powerful pesticides—poisons in reality—that accompany them but this only leads to more and more generations of multiple pesticide-resistant pests in the form of insects, fungi, and weeds necessitating higher and more frequent doses as well as constant development of hybrids, pesticide-resistant genetically modified organisms (GMOs), and pesticides. It has become tantamount to what Will Allen (2008) has termed a constant "war on bugs."

This is all central to the political ecology and economy of high modern agriculture and its global extension. Also, beyond earlier government planning, *the control of the overall system is now increasingly placed in the hands of the corporations that have the capital and research facilities to produce the manufactured inputs that the farmer, in turn, no longer has any control over; it requires larger capital investments from the farmers, and that adds to their economic precariousness; and it all has significant ecological consequences on non-targeted species as well as human health.*

A Major Critique of Industrial Agriculture

Edited by Andrew Kimball (2002a) *Fatal Harvest: The Tragedy of Industrial Agriculture* provides a comprehensive, critical look at the directions that modern agriculture is taking. Kimball refers to the *great disconnect*. Western societies began the last century as "nations of farmers with hands and minds in soil" but now led to a condition where we are "consumers who do not know where our food comes from" (2002b: 1). This is a dramatic demographic shift exemplified by the United States where in a population of well over 300 million people, there is roughly the same number of more people serving prison terms as there are farm operators—two million. In Canada where I live, out of a population of 40 million there are only 262,455 farm operators left (https://www150.statcan.gc.ca/n1/daily-quotidien/220511/dq220511a-eng.htm).

With such rapid social change away from the farm, the vast majority of people have been made ignorant in this disconnect about the realities and costs of the new food systems. As he puts it, when eating bread, they are unaware of losses of enormous amounts of topsoil, destruction of much biodiversity, and the extinguishing of most rural communities and cultures that had been supporting food production. When eating meat, they are unaware of the destruction of habitats and cruelties in the raising and slaughtering of livestock. When they eat vegetables, they are uninformed about the impacts of pesticide poisonings upon waterways, wildlife, and workers. They have not made the links in understanding the rise of cancers and the vulnerabilities to our health systems in the rise of new viruses and bacteria that can be transferred through the food chain.

While this industrial chain predominates, we are heading to a number of "fatalities" as he phrases it. Monocultural food production and artificial fertilizers in their overuse are leading to the exhaustion of topsoil. Forests are under fatal threats as more and more land is cleared for agro-industrial production. Fatal to genetic diversity is the corporate food tendency to replace hundreds of thousands of species and varieties of accumulated food options with only a few of their own selected ones that, in turn, are often genetically engineered and tied to their by-products such as pesticides. Seed saving and collective ownership among the farmers of the world are under fatal threat as the transnational companies now patent the seeds of the earth. The system is fatal to wildlife as pesticides and habitat destruction are placing many species under the threat of extinction and those that remain are greatly shrunk in numbers. In many respects, the industrial agriculture chain is fatally threatening the whole biosphere with its dependence and heavy use of fossil fuels through its mechanization.

Kimbrell writes that considerable advertising is projected by the food industry to assert the claim that this is the "only way to feed the world" in its ever-growing population. Policymakers in the crafting of international free trade agreements are also promoting this ideology.

Helena Norberg-Hodge (2002) writes of the trends toward a *global monoculture*—not just in the production of food but in the majority of economic and cultural institutions. Such a system of uniformity is clearly to the advantage of corporations, but, as she sees it, it is a major threat to the rest of us. This should be obvious in the realm of food diversity. Local ecosystems have always provided the context for embedding human social systems ultimately based on their food systems. Cultural variation reflects biological diversity. Food security was previously based on local biodiversity. Peruvian farmers will grow as many as 40 varieties of potato in the same field and that food stock will have drawn from

18 *Agriculture*

over peasant-shared 6,000 varieties with the crops being used after harvest in subtly selected different ways.

With globalization, every regional and national economy is forced into a single system, leading to the homogenization of locally adapted long-term agriculture. They become replaced by single-crop, mechanized, pesticide and chemical fertilizer-dependent, and capital and energy-intensive methods. Diversified local food security-oriented production is replaced by export monocultures designed for profit. Many thousands of local plant domesticates may disappear to be replaced by a few increasingly genetically engineered varieties produced in laboratories. Initially, the crops may be resistant to pests, but, given the extreme diversity and adaptability of potential parasites, that resistance is usually broken within five or six years and pesticides are overused until a new seed variety is found. *Farmers then become more and more dependent on seed companies.*

Much of the world's remaining cultural, agricultural, and biological diversity now lies in the global South but that is under great threat from the homogeneous, monocultural demands of economic globalization. Small farmers are being economically displaced while being drawn to the large cities that are being standardized to Northern models. Development centralizes job opportunities in the cities usually as manufacturing centers take advantage of the availability of cheap surplus labor. There in vast urban landscapes, the homogeneous pull of globalization has nothing to stop it. No matter the location, people in the cities are presented with the same inventory of general goods and food. This makes it very convenient in terms of efficiency and profit for the corporate structures of globalization and they in turn support the maxim "local crap, import good." The new migrants coming in ever-larger numbers become cut off from their home communities and the localized consumption patterns that helped to manage the biodiversity of food production.

She gives several examples of the absurd linkages generated by this global economy. Rather than producing what could be done locally, produce is brought in from ridiculous distances. Mongolia with over 25,000,000 milk-producing cattle imports butter from Germany. She suggests that for every pound of food consumed in a location, many more are generated as pounds of petroleum consumption, pollution, and waste. To Norberg-Hodge, then *the problem is economic globalization and the solution is localization.* We will return to a consideration of this in the final chapter.

Jerry Mander (2002) places emphasis on agriculture's *machine logic.* Modernity overall is dominated by that logic. It is revealed through our relationships to all sorts of machines—our cars, TVs, and computers focusing on impersonal utility. He also sees this in a general "amoralism" of the modern corporation, which is organized like a machine and acts like one. Here we have all of Earth's creatures reduced to objective potentials that relate to their profitability. They are compared and measured for their potential contribution to the economy through the metric of dollar value within the totality of a technological-industrial system. This is especially true of agriculture where once families growing diversified sets of crops fed themselves and their communities, but now, through the processes of *corporate massification,* extremely large multinational corporations have driven people from the farms into squalid living conditions in the global South.

Mander claims that 75% of the diversity in agriculture was destroyed in the 20th century. The remaining crops raised in monocultural conditions are weak and subject to insect blights, diseases, and bad weather. To counter those vulnerabilities, the overuse of chemical pesticides kills biological and genetic diversity leading to chemically laden, dead soils. As a result, constant pressures are generated to encroach on more wilderness areas with further threats to biodiversity.

Some look to genetically modified organisms as solutions to these sustainability problems. Yet these technologies function under *the same machine logic of domination and control as do the conventional forms of monoculture.* They still operate on the same principles as monocrop systems—pesticides that are designed to accompany particular crops, the limitation of the varieties of those crops, and large amounts of fertilizer, water, and chemical inputs, and ever more capital-intensive, costly forms of farming let alone the ever-increasing concentration of profits, power, and the accumulation of capital in a few corporate hands. A major threat is that of cross-pollination of non-GMO plants on a massive scale leading to some future unforeseen environmental catastrophes of disease or crop failure that could not have been predicted based on laboratory or limited field testing. Biotechnology in agriculture relies on the same ethos of maximizing machine-like usefulness out of nature.

Ron Kroese's (2002) "Industrial Agriculture's War Against Nature" traces the emergence of *the technology of high modern agriculture as highly derivative of a war economy.* The key period for this transition was World War II and its aftermath. The ideology of modern agriculture abounds with war metaphors and a conquest of nature. Many of the pesticides and their successors used in agriculture came out of that context. DDT along with other chlorinated hydrocarbon insecticides had been developed to deal with malaria in war zones. Others, including organophosphate insecticides, came out of nerve gas experiments in Nazi concentration camps. The British developed 2-4-D a highly effective herbicide during the war and that later was combined with 2–4,5-T to create Agent Orange as defoliant in Vietnam. The technologies of spraying came out of the war and the practice of aerial and other types of agricultural spraying became immediately widespread after the war because of the availability of thousands of surplus airplanes and trained pilots. The notion of spraying against pests then became widespread on suburban lawns.

Also, rising rapidly in importance was the use of artificial fertilizers. There was a tendency away from animal and mined fertilizers because of access difficulties during the war. As a result, natural gas fertilizers providing nitrogen increased eight-fold. At the time, ammonia nitrate was being produced by factories for gunpowder, and in the post-war period, the surpluses from these plants required the seeking of new markets, which were found readily in farm fertilizer markets.

In the post-war context, the notion of "farming as a way of life" was degraded, and departments of agriculture, as well as corporations, contrarily promoted it as a business and to be run as a factory. Farmers in huge numbers had to leave because they could not keep up with input costs. The notion of *a war against nature* has continued with the constant development of pesticides to combat it.

David Ehrenfield's (2002) major point is that research and development in agriculture and as a dominating tendency in biology in general has placed a huge *overemphasis on the gene as the basic unit of concern*. Organisms tend now to be seen as "packages of genes" rather than as the unique qualities of their lives in complex environments. This adds to the ideologies of reductionism that dominate agricultural research. Plants and animals are seen by many as mere units of genes that can be combined with other genes to get the characteristics desired by the human consumer from their germplasm. There is a tendency to claim "instrumental needs" for every last species and variety. These instrumental and profit-seeking tendencies draw us away from discovering the value of organisms not immediately noticed. He gives the example of mycorrhizal symbiosis then only recently discovered in soil science. Here were discovered tiny thread-like fungi that bring essential mineral nutrients to plants, and, in turn, are fed by the plants in other ways.

The Green Revolution

Having so far dealt primarily with high modernity originating in North America agriculture, what can we say about the developing world? As we shall see in the next few chapters, the mass export of North American-styled industrial agriculture is already underway, most especially in Latin America and some places in Asia. In many cases, the ultimate control of agriculture is through American corporate agribusiness giants such as Cargill, Bayer/Monsanto, and Archer Daniel Midlands.

A place where to start is the "Green Revolution" (Fitzgerald-Moore and Patel 1996) a major effort after World War II led by the Rockefeller and Ford Foundations along with the U.S. State and Agriculture Departments to bring about industrial, monocrop, high-yielding agriculture in certain developing countries such as Mexico, India, and the Philippines. They concentrated on the staple grain crops of wheat, maize, and rice. Such programs were touted as humanitarian gestures, vital to dealing with drought, crop failures, famines, and population increases in the unstable environmental and Cold War social conditions.

In the hot tropics and subtropics, several major problems challenged agronomists in that traditional grain crops when treated with chemical fertilizers such as nitrogen shoot up too quickly and collapse on their stems and when grown too closely together create uneven growth and yields with taller plants shading out lower ones. Through experiments done on wheat in Mexico led by Nobel Prize winner Norman Borlaug of the Rockefeller Foundation and involving many thousands of cross-breeding samples, it was found that short-stemmed plants could grow tightly packed and evenly with relatively more biomass in the seeds and achieve spectacular yields when heavily fertilized and irrigated. Next such experiments were tried in the Philippines with similar success regarding rice. Altogether, the advances of these high-yielding varieties (HYVs) became known as the Green Revolution and became used in developing countries Indonesia, the Philippines, Pakistan, Bangladesh, and most especially India.

The results were significant and could yield as much as four times as previously, and such countries no longer had to rely on imports from the United States, Canada, and Australia. Famines in these places were prevented but at the same time the huge population increases are in significant part due to the Green Revolution, and that globally leaves a problem that has yet to be solved.

These HYV innovations always require large and expensive inputs on the part of the farmer and have significant ecological *and* social impacts. They require massive amounts of water and nitrogen fertilizers and without them they perform worse or no better than do indigenous varieties. They also need ever-increasing amounts of biochemicals to control for ever-increasing varieties of pests that accumulate, threatening monocrop fields.

Water is crucial to HYVs and that is inevitably tied to major problems of political ecology. Previously in the Asian regions, water served more as a protective layer for the crops in question and sometimes raising fish protein among them. Now it is essential for the growing of these crops. This has necessitated canals from dammed reservoirs and electric pumps accompanying tube wells drawing from aquifers. Much has been done, especially in India, to extend such systems and as the dam, reservoir, and canal systems that have led to the displacement of scores of many millions of mainly poor peasant and tribal origin.

Another outcome is the rapid mechanization of the remaining farms, involving expensive tractors, thrashers, and electric irrigation pumps. For farmers when this happens, there are significant added expenses of fossil fuels, electricity, and repair and maintenance costs. This is obviously of great disadvantage to the small and traditional producer, especially when they are unfamiliar with credit and the institutions providing it. Social impacts go far beyond these factors..

Ecological consequences of these HYVs are profound. As with all commercialized agriculture, soils suffer especially in the subtropics where now two or three crops a year are produced in the same fields. That seriously depletes the natural fertility of the soil, because fallowing is rare now or there is a lack of replenishing winter crop rotations such as with sorghum; micronutrients are lost, and essential trace elements for more nutritious results are dangerously low. Also, with HYVs, much less straw is available to be plowed back into the soil or serve as animal feed. Salinization has been a big issue because there has been a tendency to over-irrigate and the subsequent surface evaporation leads to salt deposits in the soil eventually eliminating fertility and the abandonment of such fields. The heavy use of herbicides and pesticides has led to much reduction of the soil's capacity to resist disease. Over-compaction of the soil can result from heavy use of machinery.

Water in many areas is being overused, and this will become an even bigger issue with the full onset of climate change. Because of the ever-increasing demands for irrigation, with withdrawals of groundwater and aquifers exceeding recharge rates, water tables are lowering, resulting in higher demands for powerful pumps for tube wells, which become out of the reach of less prosperous farmers.

Political Motives for the Green Revolution and Other Global Agricultural Policies

While undoubtedly having some humanitarian motives, observers such as Engdahl (2007) have suggested that the Green Revolution and subsequent attempts and successes to globalize agriculture are, at its heart, an American attempt through corporate control to dominate world agriculture. It is hard on the basis of their evidence to deny this as having significance. Engdahl (2007) takes us back to the immediate postwar period and the Rockefeller Foundation when the United States was creating a new kind of empire. Land would not be occupied by foreign powers, but former colonies and less developed countries would be controlled through economic policies to the advantage of the United States that dominated the global economy by about a whopping 50% at the time. A new kind of imperialism was created through the dominance of corporations backed by a powerful military supplemented by an atomic arsenal. This empire could be managed without physically occupying other nations' territories and also through the international rule of laws and institutions that the United States itself had established such as the World Bank and the International Monetary Fund.

Nelson Rockefeller of the oil-rich Standard Oil family had succeeded in extending American interests and new markets in Latin America and saw that oil and agriculture could be well combined in efforts to promote a highly efficient and large-scale business-styled agriculture. Rockefeller was also working with the gigantic seed and grain marketing company Cargill of Minneapolis in schemes to intensify Brazilian agriculture in the direction of hybrid grains, soy, and large-scale hog production using techniques that were quite advanced. There was an emphasis on the use of synthetic fertilizers derived from oil in these deliberations. Rockefeller frequently made use of the motive of "national security" and he had continuing major influence on the U.S. State Department where he had been Undersecretary for Latin American Affairs during World

War II. What Rockefeller was doing in Latin America was to get a head start and be a prime innovator for a new kind of revolution in world food production, and the Rockefeller Foundation even gave the process the new name—"agribusiness"—generated by an industrial model and fostered through a free market ideology. Altogether the interests in this scenario were intended to "take over the control of the basic daily necessities of the world's population" (Engdahl 2007: 117).

Behind all of this was a glut of nitrogen fertilizer, and since that industry was an offshoot of Standard Oil, other companies were interested in greatly extending their global markets. That generates links to Norman Borlaug, the founder of the Green Revolution, who worked through the Rockefeller Foundation's support of agricultural research in northwestern Mexico and his role in kick-starting the development of HYVs. This was the beginning of the now hegemonic control over global agriculture by the corporate sector. As the Green Revolution extended into areas beyond Mexico, it focused on particular regions, such as the Punjab region in India because by targeting farmers there that were already large and wealthy and oriented toward capitalist and industrial approaches, they as early adopters, capable and through their credit capabilities, would spread the innovations.

Among the earliest of those associated with Rockefeller interests was Henry Kissinger who served as a director of some Rockefeller Foundation projects and later came to hold very powerful positions as National Security Advisor and Secretary of State in the Nixon and Ford presidencies—and in the latter Rockefeller served as Vice-President. Kissinger was notorious for considering food as a weapon equal to military strength in asserting American agendas during those regimes that overwhelmingly favored agribusiness interests to those of farmers.

The United States during the early 20th century had developed its economy by nationally integrating both its industrial and agricultural food products and placing high tariffs on imports to protect and solidify both sectors. With the industrialization of its agriculture, it strove and has for a long time succeeded in being the breadbasket of the world, especially after World War II. As it has already been indicated, that still requires farmers to purchase massive amounts of input products—seeds, fertilizers, pesticides—that come from an industrial sector run by corporations. In the post-war period, these corporations with the help of the U.S. government started to export such inputs across national boundaries through American aid programs in developing war-torn Europe and Japan and in other areas such as Brazil and the cone of Latin America that provided feed for large-scale production of meat. Then with the Green Revolution that reach has been even further. At the same time, American products such as high fructose corn syrup and soy oils began to replace cane sugar and tropical vegetable oils were major ingredients in the rapidly expanding food processing industries. They too were widely exported. So, there is no question that the shape of the global food regime has been determined by American policies and secondarily by its European allies—former colonial masters.

The United States during the 1930s had followed a neo-mercantilist approach to build up its agricultural strength by banning agricultural imports and developing price support systems, and this led to major surpluses. These were then converted into international aid programs, and the global distribution of American food further reinforced the corporate positions of American grain traders such as Cargill, and the technologies of farm input soon followed as exports. Joint ventures with foreign countries and America started to integrate international commodity chains involving ever-more specialized agricultural sectors across national boundaries, thus moving toward a model of full global interconnection.

By 1994 the agricultural power of the United States was dominant globally. Besides its massive industrial clout American exports were highly dominating—36% of wheat exported, 64% of corn, barley, sorghum, and oats, 40% of soybeans, even 17% of rice, and 33% of cotton. Those that have been most strengthened by all of this though have not especially been American farmers but American corporations. Brewster Kneen (1995) documents the amazing lobbying power of Cargill. It has gained astonishingly high subsidies to bolster its overseas businesses and has a frequent revolving door with officials going to the government and back to bolster trade agreements in favor of the company. In the United States itself, the food industry is actually the largest manufacturing sector through food processing and thus a very powerful political economic entity.

For the global South, the strategy of the Northern developed countries through their neo-colonial strategies has led to the undercutting of local farmers by flooding markets with products from the North. At the same time in these same countries, it has been constantly reinforcing Green Revolution ideas that create dependence on costly inputs and open the door for American and European agribusiness. This creates vulnerability in *both* local food security and food sovereignty and causes situations of poor nutrition with a growing emphasis on processed foods, especially in the massively growing cities in the developing world. The twin strategies from this globalization also tend to exacerbate existing inequalities. For instance, Brazil, one of the world's top and rising agricultural producers, has circumstances where less than 1% of its population owns 44% of fertile land while 32,000,000 people are labeled as being destitute.

While the big grain trading, agrichemical, fertilizer, and seed companies exercise their power on the global commodities markets, they also, through free trade agreements and the operations of the World Trade Organization, exert powerful lobbies to liberalize trade. They question and attack sovereign countries' regulations such as supply-side management procedures that control the amounts of particular commodities and thus try to get fair prices such as dairy, egg, and poultry for the farmer producers. Major food processing companies such as General Foods, Kraft, PepsiCo, and Kellogg's have been able to break down foreign ownership limiting rules and set up plants in countries such as Mexico and now dominate local industries. Altogether, such operations, aided and abetted by neo-colonial governments and through the World Trade Organization, came to create a neoliberal economic order globally. The colossal irony, hypocrisy, and major injustice are that the United States and its allies in the developed world, having achieved dominance of global agriculture through the use of protective tariffs, have been demanding that nations in the Global South drop all of theirs in the name of "free trade."

Value Chains

With globalization, agricultural food chains, the links from the growing of the food at the farm level to the consumer's plate often thousands of miles distant, have become quite complex. Most winter fruit and vegetables as well as specialty crops such as coffee and flowers come from distant sources to North America and Europe. Many middlemen are involved, and in cases such as with coffee, there will be significant value-added dimensions along the way with packaging, grinding, and with transformations into instant versions, or particular blends. For North America, we are referring to products primarily sourced in Central America. The markets are controlled on the demand side rather than the supply side (Dowdall and Klotz 2014). This situation places the large agribusiness food processors and distributors in by far the most powerful positions compared to the farmers in the distant source. The major distributors, though, do have certain market pressures placed on them—the North American consumer demands the cheapest and steadiest prices possible and guarantees of stable or improved quality in the condition of the products. These stipulations are central to the agendas of highly competitive larger-scaled grocery chains as semi-monopolies in affluent northern countries.

What happens is that most of the pressure and negative impact are diverted downwards to the distant developing countries. The buyers and suppliers in producing countries such as Guatemala struggle to win and keep the lucrative large-scale contracts with the American distributors, who, in turn, sell to the mega-corporations that process raw products and sell them to supermarket chains. Tight and very uniform quality standards are imposed on the many thousands of peasant farmers at the bottom of the value chain. They are always expected to bear the brunt of any downward market fluctuations on their commodities since the bulk buyers are further up on the chain of economic and political power. To manage quality control, the farmers are subject to pressures to buy seed and any other required inputs from corporations rather than using their traditional methods of using few inputs and saving seeds and sharing them with fellow peasants. Most of all, it becomes unavoidable for them to eventually buy expensive synthetic fertilizers to ensure large yields and to purchase toxic pesticides to control and guarantee harvests. Altogether this all places these farmers on a "treadmill" parallel to the ones that North American farmers and those targeted for the Green Revolution have been on. Cost-squeeze pressures drive many out of farming and force the remainders to get bigger. Then more people are forced to move to large urban centers where high unemployment and low wages face them. Their domestic food security is threatened, and they find themselves flooded with cheap corn and wheat from the dominating countries where the value chains originated. In the Central American case, this even leads to mass attempts to emigrate to the United States where they assume that industrial jobs are waiting for them. The social, economic, and environmental impacts of these trends are hugely negative.

Foreign Land Grabs

Land grabs or euphemistically "large-scale, land acquisitions" are also beginning to threaten both food security and especially food sovereignty (i.e., the capacity to manage a country's own food sources) among local peoples in the developing world. These purchases or leases are those of 1000 hectares or more and done by corporations seeking to establish palm tree plantations or other large food or fiber producing operations in other countries. They sometimes do this to produce biofuel ethanol or for food destined for their home countries not local consumption. The World Watch Institute (2015, https://www.commondreams.org/newswire/2015/10/06/land-grabbing-grows-agricultural-resources-dwindle) reports that more than 36 million hectares, about the size of Japan, had already been purchased, and another 15 million were under negotiation.

There are three broad parties involved—the foreign buyers, the national governments of the affected countries, and the local people, who had for their own subsistence, previously been dependent upon the land now purchased by others. Rarely are the local people consulted or compensated for their losses. The impacts upon them can be extreme, as they may be totally displaced, losing a subsistence base upon which they had developed farming techniques that had ensured long-term sustainability. African pastoral nomads with traditional transhumance patterns have been especially affected. In Africa, the major target of land grabs, less than 5% of the total land mass is under legal title but traditional methods of land tenure, especially those under commons jurisdiction, are very effective in maintaining order and distribution.

The second party, the affected nation-state, can gain in the form of sales or rental income, which they may then use as revenues for the operations of government and get to claim that the new industries are positive developments for the country. Unfortunately, in some cases, the standards of government integrity may be low, and the involved officials take bribes or distribute all of the profit among themselves. The investors often represent countries that are short on land and food production themselves, so the food and water (converted into food-in-kind) are exported from the donor nation to feed the buyers' own countrymen. Countries engaging in this practice include China, South Korea, Japan, Saudi

Arabia, and some other Gulf States—all rich in capital. The countries most affected include Ethiopia, Brazil, Guatemala, the Democratic Republic of the Congo, Mozambique, Pakistan, and South Africa—among the least developed.

Between 2010 and 2013, China had bought 7.5 million acres mainly in Africa, and Saudi Arabia had spent over $800,000,000 on overseas farms (MacVittie 2013: 26). The long-term environmental consequences of these modern enclosures include the loss of much water since the new farming procedures are now intensively drawing from aquifers and surface waters. The general problems outlined for North American and Green Revolution agriculture apply in these cases. Most people are displaced with only a few low-paying jobs available for local people. It has generated conflict, in some cases armed, and can be seen as an extreme infringement of social justice (see Vangerdarten et al. 2016; Krieger and Leroch 2016).

Conclusions

Contemporary North American agriculture and its global extensions through various green revolution innovations are a little over one hundred years old. They are based on high modernist and capitalist principles. Advocates hold assumptions about the supposed superiority of industrial capitalism in agriculture—seeing value through constant growth in the accumulation of capital; the sovereignty of the market; the sanctity of private property; and ever-increasing efficiencies in the extraction and manufacture of commodities. The system *has* achieved successes through the maximum application of science and technology directed to farming as it had in other domains of production. Yet such principles came later to agriculture because it was much more difficult to apply them directly to Nature in all her complexity. Still, through persistent investment, research, and development applying principles of agricultural engineering and plant science, the emerging, mechanized, monocrop-based system has prevailed and has seemingly tamed or subdued many aspects of Nature. But that is really just for now—we need to constantly remind ourselves. Through ever-more powerful machines, constantly reducing the need for labor, and extending the amount of land to be cropped, tractors, swathers, combines, sprayers, external and portable systems of irrigation, and other devices along with the constant development of high-yielding hybrid varieties of plants accompanied by synthetic fertilizers, especially nitrogen and fast action, lethal pesticides, high yields have been achieved.

Taken to the most extreme of industrialization has been the highly modern techniques for producing animal protein—now largely an industrial product. Intensive livestock operations or confined animal feeding operations (CAFOs) involve minimal labor and use of mechanized ways of delivering feed and water and removing large amounts of fecal and urinary waste. With respect to hogs and chicken, they produce massive amounts of meat, faster and at much lower per unit costs than could have been imaginable 50 years ago—although there have been huge ethical, health, and environmental consequences. American agriculture became the most dominant and specialized in the world and it served national purposes well for a while with large numbers of laborers being freed from or displaced off the farm especially at the times surrounding the two world wars, thus helping to build up the complementary industrial might of that nation. These dynamics were also true of other developed countries such as Australia, Canada, and in Europe.

Most of these trends and the high productivity of agriculture were directed by the rise of agribusiness corporations in their positions of wealth and ever-growing power in their capacities to dominate the accumulation of capital and profit for their sectors in the food industry. These agribusinesses include machinery manufacturers such as John Deere, Case, and New Holland, grain handling companies such as Cargill, Bunge, and Archer Daniel Midlands, fertilizer companies such as Nutrien, Yara, and Mosaic, pesticide and seed companies such as Bayer/Monsanto and DuPont, and vertically integrated giants such as Tyson, Smithfields, and Perdue Poultry in the meat industry. Then there are the major food processors—Heinz, PepsiCo, Nestlé, General Mills—that constitute the largest sector of contemporary food processing. All these companies are fierce in their competition and many are oriented toward both horizontal and vertical integration (Kneen 1995). Still, there has been a dangerous tendency toward monopoly capitalism in the tendency to eliminate or absorb thousands of smaller companies and there are already worrisome forms of hegemonic power. Strategies used by advocates, especially through the media, are oriented toward promoting the advantage of cheap consumer goods through its production techniques and to extending them to a global market as supposedly the "only way to feed a hungry and ever-growing world." The uninformed, urban consumer through the "great disconnect" no longer knows about the realities of growing food and rural life to dispute this view and tends to complacently accept such claims.

Meanwhile, many more farmers are constantly leaving their occupations because of stresses and economic failures in keeping up with these Social Darwinist trends in agriculture. The policies of departments of agriculture in developed countries constantly promote these industrial-styled efficiencies with the ideologies that farming could no longer be considered a way of life but only as businesses and that operators had to "get big or get out." A cost-squeeze treadmill has been generated, and the remaining farmers end up renting or buying up large amounts of land to sell ever more commodities just to try to stay even. In American and European cases, subsidies have propped up certain categories of their farmers leading to flooding of the market with cheap commodities that benefit the food processing and meat industries. European and American free trade advocates hypocritically demand that other countries eliminate their agricultural tariffs. Yet the farmers continue to suffer economically and keep falling off the treadmill. Many farmers as primary producers have now become similar in position to proletarian workers or feudal serfs through their contractual relationships with corporations depending on their particular type of specialized farming.

The impacts on rural communities have been profound. With massive depopulation, North American communities have frequently become rural pockets of poverty populated by ever-growing portions of the elderly. The patterns of face-to-face assistance and cooperation that may have been built up during pioneering phases are evaporating. The services of schools, hospitals, railway stations, grain elevators, anchoring friendly settlements with their churches, stores, hotels, and coffee rows are rapidly disappearing. The senses of abandonment and neglect are deep and find expressions sometimes in populist politics that some might say miss the real targets of their woes. In the United States, there is not much left in the way of a surplus population to be absorbed through urban migration.

On the other side of the disappearing and struggling farmers are the powerful institutions of society that include the agro-corporations, government departments of agriculture, and university colleges of agriculture. With the decline of direct public funding for agricultural research, more is left in the control of capital-rich corporations that enter into public-private partnerships through funding and the selection of research topics with revenue-scarce universities as well as governments. Beyond these significant connections, governments themselves actively promote these industries and their innovations. American administrations, in particular, are noted for revolving-door appointments in their Department of Agriculture and Environmental Protection Agency between industry and public services.

The upshot of all of this is that all these same entities are actively and very aggressively promoting these types of agriculture globally and because of hegemonic power have been able to convince both the public and policymakers that, once again, this approach is "the only way to feed a hungry world." The Green Revolution has been a clear example of this and the globalization of multiple food chains. There have been huge social consequences from these actions. Some of the worrisome ones have included large-scale land grabs reminiscent of the enclosures from previous centuries, displacements in the hundreds of millions from rural homelands, cost-squeeze pressures, and the hazards of being on the losing end of the long value chains in which they are engaged. The cost-squeeze pressures there place farmers on an equivalent and maybe even more severe treadmill to their colleagues to the North as they too become ever more dependent upon expensive chemical fertilizers and pesticides. Inequalities abound, and many are forced into becoming landless laborers and when they migrate to huge cities in the developing world, they are more apt to end up living in large slum shantytowns than they are to be successfully absorbed into a responsive manufacturing economy.

Environmental concerns about these forms of agriculture loom large in both the developed North and the developing South. Water from dammed reservoirs and aquifers is dangerously depleted in many parts of the globe as hydrological cycles are being distorted and perils will increase with the onslaught of the complexities of climate change. Biodiversity, right down to the amphibian, insect, and microorganism levels, is under very serious threat for a multitude of reasons including the overuse of pesticides, the maximization of land use, and the massive, habitat destroying clearing of forests for agriculture. The last also removes major carbon sinks and disrupts local weather patterns.

Agriculture plays a major role in greenhouse gas releases—methane from large-scale livestock operations has 20 times the per unit impact as does carbon dioxide. Carbon dioxide released through large-scale mechanization of agriculture and the long distances that food commodities travel as well as processing account for a big proportion of greenhouse gases.

The success of this high modernist agriculture in its global context is heavily dependent on the oil industry. Some (Jackson 2010) have calculated that for every calorie of food produced 10 calories of alternative energy are required—thus a negative ratio of 10 to 1. Soils have been and are being seriously degraded. Organic nutrient replacement through animal manure and human waste recycling has been all but eliminated in many places of the world. Substituting for that has been the huge reliance on mixtures of synthetic fertilizers of nitrogen, phosphorus, and potassium that do not contain the trace elements necessary for healthy soils let alone healthy nutrition. Since such fertilizers are water-soluble, run-off especially of nitrogen has caused many ecological problems—such as the eutrophication of river mouths and large swaths of coastal waters, and cancer-causing nitrates threaten human drinking water sources. Pesticides appear to be killing off many microorganisms within soils that may have highly significant roles in the micro-dynamics of nutrient delivery to plants. Above all, globally, we are faced with massive problems of soil erosion as more vegetative cover is removed from fields and fields are left open for continuous mono-cropping. Even an advanced agricultural society such as the state of Iowa has lost one-half of its topsoil depth in a little over 150 years of settlement (Pimental and Burgess 2013).

There is enough food or at least grain crops to feed a hungry world but much of it has been diverted to feeding livestock, and several billion people remain insecure with regard to food. Even the seemingly affluent can be viewed at risk with poor nutrition and obesity based on processed foods packed with surplus commodities such as high fructose corn sugar. Long value chains also create significant health challenges such as salmonella outbreaks and bird flu.

Complexity theorists refer to *path dependencies* and *phase transitions*. Path dependencies occur when a series of trends and actions lead to their shifts and dominating consistencies in what continues to transpire pushing out older trends. Phase transitions occur when major changes (sometimes irreversible) occur to the systems under examination (Ervin 2015). Path dependencies regarding North American agriculture started in earnest around World War I and have occurred unabated leading to a major phase transition of mechanized, monocrop, industrial, and confined animal feeding operation types of agriculture that are unlikely to be reversed. Although future events of ever-increasing chaos and instability might generate new phase transitions. That is where climate change is likely to come into play. As a thought experiment—consider what if agronomists were aware of what we know from ecological science, agroecology, agroforestry, permaculture, and other scientific ways of growing food, could have a phase transition taken another direction? We will never know but we could still ask that question now of the developing world where hundreds of millions of peasant farmers feed the bulk of their populations. We will come back to these questions and possible solutions in Chapters 5 and 10.

3 Field Crops
Grains and Soy

This chapter deals with the production of plant food in open-field, monocultural systems. Case studies and general discussions in this chapter are largely devoted to the cereal crop of corn, which is among the major staple foods feeding humanity, and with soy, a bean, which though can be linked with corn as the major industrial food crops going far beyond direct human consumption—approaches taken to their production are really dictated to serve the agribusiness interests. Corn and soy play huge roles as animal feed for the complicated food value chains in the global meat industry.

Corn

An excellent place to start is Michael Pollan's (2006) *The Omnivore's Dilemma: A Natural History of Four Meals*. Pollan informs us that corn is the main ingredient of a Mexican's anatomy, which is understandable given its origins in Central America and its position as a staple in the tortilla-dominated diet there. As he points out, this is also true of many other nationalities, not the least Americans, where biologists can determine the dominating source for carbon atoms, the essential building block of all life forms as coming from corn because of the huge role it has come to play in processed food and in the overwhelming use in animal feed.

Corn, domesticated by the indigenous peoples of the Americas, proved to be extremely useful to European settlers. With the early help of Native Americans, they were shown how the plant was adaptable to just about every micro-climatic condition since wheat varieties from the Old World did not manage well at first. It was favored by American pioneers because of its much higher yields per planted seed (about three times that of wheat). It was ready to eat as a vegetable; easy to store; was a source for flour; useful for animal feed especially hogs; cobs could be converted to fuel; and as a distilled beverage it was a source of inebriating solace on the frontier. Once domestic needs were met, it was easy to transport and sell in town markets.

Because of corn's dependence on human interference in reproduction, Native Americans had already developed thousands of crossbred varieties with unique and highly useful qualities. When professional and consistent hybrid crop breeding became the norm in the 20th century, corn was ideal for that process. It was constantly being made variable in adaptation to human changing needs. It was favored for industrial innovation in food production and processing. It became upright, stiff, and extremely tightly packed in conformity to industrial machine-driven harvesting. It became ever more numerous in its proliferation so that as many as 30,000 separate plants could be grown in a single acre with ever-increasing yields. It could adapt to industrial chemicals especially artificial nitrogen as its food and be tolerant of poisons that eliminated its competitors. It was even to become a form of intellectual property through its hybrid production because, after two generations, its seeds would not reproduce, and farmers had to keep buying ever-new varieties of seed stock. This returned large amounts of profit to corporations that had the facilities to manufacture these hybrids. This led to farmers becoming increasingly dependent on the corporations for fertilizers and pesticides as well. Eventually, "it brought the whole American food chain with it" (Pollan 2006: 31).

In his exploration of corn and tracing its domination of the American food chain, Pollan begins at a contemporary Iowa farm. His host's grandfather was able to only get a tenth of the yield (20 bushels an acre) as today with widely spaced self-pollinating seeds in clusters rather than tight rows, thus about the same yield as Native American varieties. Then commercial hybrid corn types first came into play during the 1930s and the farmer's father was getting 70 to 80 bushels per acre and in the current generation has led to the possibility of harvesting 200 bushels an acre.

In earlier times, there was also a tendency to devote surpluses to mixed hog and grain operations. That basic mixed-grain/livestock raising form has eventually all but disappeared as there was growing pressure through government policies and from agribusiness to plant virtually every acre in corn. A very significant part of the changes was manifested in the massive surpluses of corn that then could be devoted to feeding hogs now in confined animal feeding operations (CAFOS) as separate intensive operations also of an industrial nature. American farm policy has long been connected to ever-increasing yields of corn even when the prices start to go way down. Farmers of this crop have been caught in a cycle of expanding their acreages and yields to keep up with their costs of production but lower returns. Besides the end of mixed hog grain farms, the surplus corn began to be used in large quantities to the feeding of chickens in CAFOs and to cattle in feedlots. By the 1980s, corn was "king" in Iowa.

Another important impetus for the industrial development of corn was the conversion of World War II munitions plants with their surpluses of ammonium nitrate to the making of chemical nitrogen fertilizers. More than any other plant, corn requires huge amounts of nitrogen, so this was especially timely. It also makes corn the exemplar of industrial agriculture. Midwestern American soils could not have supported such productivity on their own for much longer. The underlying chemical method—the Haber-Bosch process for artificially fixing nitrogen—is probably one of the very most important inventions of the 20th century. This availability of artificial fertilizer also reinforces the monocrop character of Iowa corn growing—the farmer no longer has to spend half the growing time covering his fields with crops such as alfalfa or beans to help recondition the soils. *He can depend on oil to do that for him*—it takes between a quarter and a third of a gallon of oil to grow a bushel of corn and about 50 gallons per acre. Another way of looking at it is that it takes much more than one calorie of fuel to produce a calorie of food. Even before the introduction of chemical fertilizers, the ratio had been two calories of corn per calorie of fossil fuel—now it is ten fossil fuel equivalents to one of corn.

Ecologically, this is extremely inefficient and contributes significantly to carbon emissions. Previously, the growing of corn was in effect a free gift from the sun but the service was a lot slower and much less abundant. "In the factory time is money and yield is everything" (Pollan 2006: 46). But as he points out, factory processes inevitably bring pollution and the corn, while "eating" fossil fuel products, ends up being fed much more than it needs. The recommended amount is a hundred pounds of nitrogen fertilizer per acre but farmers frequently double that just to be sure. With the heavy rains, a large amount makes it into the water table making precautions necessary in protecting water sources. Some escapes into the air as nitrous oxide significantly contributing to global warming, and more leaks into drainage ditches, streams, and rivers making its way for instance to the city of Des Moines where "blue baby alerts" are made during the spring. Here, warnings are made about giving babies water from taps since nitrates in the bloodstream can poison them so that red blood cells that carry oxygen to the brain are seriously impaired. Eventually, the nitrates make their way through the Mississippi River system where the algae that thrive on it create a huge coastal dead zone by smothering through eutrophication the oxygen that would normally go to supporting other marine life.

Oddly enough, at the time of Pollan's writing, the market price for corn was about a dollar less than the cost to the farmer of growing it, which makes no sense. Yet just about everybody else—the consumer, the livestock producer, the food processor and manufacturer, and the corporations that variously supply the farmer or handle his crops—benefits enormously from such low prices. This originates with the strange reversal of agricultural policies generated by successive American governments beginning with the Nixon administration. During the previous New Deal era, governments determined a target price for corn based on the costs of production; when prices fell below it, farmers were encouraged to store grain and reduce production. They could take out loans based on corn as collateral and then later sell the corn when prices were higher or keep the loans in exchange for giving the corn up to a national granary program. The government might then use the corn in its overseas aid programs. Farmers were also encouraged by various New Deal programs, including payments, to take fields out of production for a while to prevent soil erosion and overproduction.

Pressures from Wall Street and big business long objected to these support policies, and their focus was trying to get corn and other grains subject to purely market pressures of supply and demand. This was done in ideological opposition to the then-strong lobbying and political capacities of farmer organizations. Then along came Earl Butz, an agricultural economist and then a board member of a giant food-processing corporation, who became the transformative Secretary of Agriculture during the Nixon government. In 1973, he took American food policy on a 180-degree turn. The early 1970s saw food prices reach a high because of crop failures in Russia, which was favorable for farmers. But consumers were hurting. Butz started programs to push basic food prices down and greatly increase supplies. He encouraged farmers to get ever larger in their operations—"adapt or die"—and plant every acre possible while identifying themselves as "agribusinessmen." He eliminated the National Granary and government purchases and loans and replaced them with direct payments to farmers. There was no longer a floor under the price of grain with the prices encouraged to go low and farmers encouraged to sell at any price.

In many cases, the farmer ended up selling at a loss. What replaced the New Deal was a new set of subsidies in which the farmer is paid an advance before even putting in crop of each acre intended for corn with an equal amount coming later. In the 2007 film *King Corn: You Are What You Eat* the authors rent a single acre of land from an Iowa farmer and follow it through a cycle and where potentially the harvest will go. Their expenses were $349.00 and they are paid $330.00—so a loss of almost $20.00. They reported $28.00 in federal subsidies and then several others, such as "deficiency" payments are mentioned, although they do not enumerate them. In Pollan's year, 2005, it cost $2.50 to grow a bushel of corn and farmers received $1.45 at the grain elevator (Pollan 2006: 53). The collective outcome is that farmers grow as much corn as they possibly can, buying up or renting as much land as they can, eliminating family gardens, other crops and livestock, woodlots, and margins between farms. The year the film *King Corn* was made, Iowa produced two trillion bushels of corn.

Why do farmers keep doing this? It seems that farmers and their organizations have switched their philosophies to adjust to this bizarre reality by touting the American dream rhetoric of being businessmen and engaging in free markets and free enterprise. Yet hundreds of thousands of farmers go deeply into debt and thousands declare bankruptcy. Furthermore, the reason that they continue in this trap is that they must have cash flows to support their minimal standards of living, pay bills, and service debts—a self-reinforcing rut. They have to sell more and more corn even as the input costs keep going up including more liquid nitrogen to grow greater yields. The growing of these crops in such large quantities impoverishes farmers in the United States and the countries where it is exported. For the farmer in Iowa, corn and soybeans are the only crops that will be accepted in the local market—the town grain elevators. Farms are not like factories—they cannot lay off workers since they are family members who must be taken care of.

These federal agricultural policies have been heavily influenced by the major agribusiness corporations and they are the real beneficiaries of the subsidies. American taxpayers really have been subsidizing *them* in the scores of billions of dollars because the overproduction creates extremely low market prices. Fundamentally what it does is provide enormous amounts of cheap carbohydrate energy available to corporate food and beverage companies such as Coca-Cola, McDonalds, and General Mills, and high profits for seed and herbicide companies such as Bayer/Monsanto, farm implement dealers such as John Deere, and fertilizer companies such as Mosaic. The biggest players in the purchasing and marketing are the same ones we will see again when we examine soy—being Bayer/Monsanto, Cargill, and Archer Daniels Midland (ADM), who control the seed and product purchasing, transportation, marketing, and even processing of corn as well as being the most significant influences on federal farm policies. The primary destination for corn is confined animal feeding operations (CAFOs). Three out of every five kernels of corn go there—for chicken, hogs, and cattle.

The amount of corn directly eaten as white corn (rather than the industrial and feedlot corn) is a very tiny proportion of the harvest. Here we are talking about it as on the cob or in baked products such as tortillas, tacos, cornbread, or chips. Still, the yellow or "number 2" corn that we have been discussing—that which is massively produced and already makes its way into our bodies and has other ways of being processed to the benefit of corporations such as General Mills, Coca-Cola, and McDonald's. About two-fifths of this corn goes into what are called wet mills. This is in contrast to a dry milling process that grinds white corn into flour. Wet mills are similar to oil refineries in that the original product, in this case carbohydrates, can be broken down into many other chemical products—ethanol for alcohol or use in cars, glucose, fructose, citric and lactic acid, mannitol, xanthan gum, MSG, and others. In the wet milling process, according to Pollan, it is the mechanical, industrial equivalent of digestion—breaking a complex carbohydrate into simple molecules. Probably the most significant product of milling processes is high-fructose corn syrup (HFCS) composed of 55% fructose and 45% glucose and tastes as sweet as sucrose. It has virtually replaced cane sugar or sugar beet sources for sweeteners in production because of much lower costs and huge surpluses. Beyond the sweetener function it can be further processed into adhesives, coatings, plastics, thickeners, gels, and other uses.

Pollan suggests a remarkable revolution in the adaptation to food to humans and our adaptation to it with the manufactured dimensions of corn along with its sister product, soy. Referring to the machinery involved in the processing aspects of corn he writes

> Step back and look for a moment and behold this great, intricately piped stainless steel beast: This is the supremely adapted creature that has evolved

to help eat the vast surplus biomass coming off America's farms, efficiently digesting the millions of bushels of corn fed to it each day by the trainload. Go around the back of this beast and you will see a hundred different spigots, large and small, filling tanker cars of other trains of with HFCS, ethanol, syrups, starches, and food additives of every description. The question now is, who or what (besides our cars) is going to consume and digest all this fractionated biomass—the sugars and starches, the alcohols and acids, the emulsifiers and stabilizers and viscosity control agents? This is where we come in. It takes a certain kind of eater—an industrial eater—to consume these fractions of corn, and we are, or have evolved into, *that* supremely adapted creature: the eater of processed food.

(Pollan 2006: 90)

Food preservation was enhanced by the processes of canning and freezing but also with the introduction of a type of manufacturing that now transcends those basic types. Beyond freezing, canning, fresh produce, and meat, the vast majority of long-lasting processed foods on supermarket shelves will contain mixtures of these corn-derived ingredients as well as many concocted from soy. The research and development for such products becomes a highly secretive and competitive corporate domain where such products are constantly devised. Pollan gives examples from the processed cereal domains where a four-dollar box of cereal is designed from about four cents of corn or some other grain and combined with corn starch, corn sweetener, and ingredients that generate pleasing colors and shapes, along with added vitamins and minerals to give the veneer of healthiness. This is all followed by an expensive marketing campaign to sell the product.

As he points out, capitalist food production faces a set of realities. On the consumption end, there is only so much food that a human can eat in a single year—a biological fact. Related to that is that fundamentally the growth in a food market then would be limited by population growth, which in the American case is about 1% per year. Then on the other end—the production end—biology can potentially raise the costs of the raw product of food through scarcity brought about by disease and crop failure. The challenge then is to find ways of procuring more ways of capturing the food expenditures of the consumer, and that is where the processed food industry comes in with remarkable success by designing ever-new more expensive products than the raw sources. Processing food also adds many months to the shelf lives of the products, and in these value-added strategies, it allows the companies to continually seek new markets globally. The farmer gets the short end of this profit chain. For every dollar spent on corn sweetener, the farmer gets four cents, and the corporations get the rest.

One of the biggest sources of profit and at the same time adding to major health concerns is the predominance of the soda drink industry. Corn whiskey in its proliferations used to be a major concern in the United States, now it is soda pop based on massive amounts of high fructose corn syrup (HFCS). As he points out, three out of five Americans are overweight and children born after 2000 supposedly have a one in three chance of developing diabetes. Globally, this is a factor as well with one billion now being considered obese. Mechanization and less physical exercise constitute one set of factors, but calories have been substantially added to our diets. Yet most of that is hidden in the processed foods. Two hundred calories have been added on average for every person each day. Consumption of added sugars has gone up annually from 128 pounds per person to 158. It is everywhere essentially—crackers, ketchup, mustard, cereals, hamburger buns, drinks, and many other products. Soft drinks appear to have been the major place where the switch to HFCS occurred with their adoption by Coca-Cola and Pepsi in 1984 due to the market advantage of using it in place of other sources of sweeteners.

This system has been able to take advantage of the human evolution toward "elastic" appetites for sweet or energy-dense foods as a means of putting on extra reserves of fat to manage "lean" times. Yet in nature, humans rarely encountered energy-dense foods that can match the corn-based fructose in a supersized soft drink or in a chicken nugget. Processed food is an excellent strategy for getting people to consume more food and for growing profit. Given the ever-lowering price of calories found in these sugars and fats, the further down on a socioeconomic scale, the more vulnerable people are to this phenomenon and its health consequences. And in spite of frequent health warnings about the consequences of this type of diet, successive American presidents continue to sign farm bills that subsidize the industries that profit from cheap corn.

In keeping with Michael Pollan's tendency to suggest that plants in a certain sense domesticate we Homo sapiens just as much as we have domesticated them. He suggests that corn or *Zea mays* itself has been the big winner.

> Corn's triumph is the direct result of its overproduction and that has been a disaster for the people who grow it. Growing corn and nothing but corn has exacted a toll on the farmer's soil, the quality of the local water and the overall health of his community, the biodiversity of his landscape, and the health of all the creatures living on or downstream from it. And not only those creatures, for cheap corn has also changed, and much for the worse, the lives of several billion food animals, animals that would not be living on factory farms if not for the ocean of corn on which these animal cities float.
>
> (Pollan 2006:218)

Pollan's writing on corn truly sets the stage for an understanding of the global industrial food chain and its power with largely negative impacts upon the rest of the world.

NAFTA, Corn, Mexico, and the United States

Corn has been the mainstay of American pre-eminence in global agriculture, yet Mexico was the source of its innovation and development among indigenous peoples. As Soleri and Cleveland (2006) point out, the corn regimes of Mexico and the United States could not be more different. Ironically, in Mexico the motherland of corn, the culture, society, and economics of it are under threat through the domination of American corporate control and the manipulation of global trade policies. Corn, unfortunately, serves as an exemplar of the political economic, social, and environmental dangers of those who have the most power—i.e., American agribusiness, transnational corporations, the American nation-state, and neoliberal, globalizing interests among the wealthy in Mexico itself.

About 99% of maize varieties in the United States are modern varieties and practically all are now genetically engineered (about 90%) historically arising out of scientific hybrid seed breeding processes in the 1930s. Native Americans as traditional growers or as with Euro-American farmer varieties on organic farms constitute less than 1% grown in the United States. In contrast in Mexico only 21% is of the modern hybrid variety and not yet legally genetically engineered. The types of GMOS (genetically modified organisms) grown in the United States are Bt modified (containing an ingredient that kills corn borer larvae) or herbicide resistant as with pesticides containing glyphosate such as Roundup, or a combination of both characteristics. In contrast, only some very limited experimentation with GMOs is permitted in Mexico at research stations although it was once banned entirely. In Oaxaca where Soleri and Cleveland (2006) have been conducting considerable research, only 10% are of the modern but non-GMO varieties. Mexican farmer varieties of which there are well over 10,000 include landraces (traditional types that have grown in isolation and specifically adapted to those particular, local environments and with a huge amount of genetic variation within each type) or that have occurred by natural selection. Farming households in Mexico typically consumed a portion of the corn as part of their own diet and sold surpluses to feed the vast majority in towns and cities. The corn that is grown in Mexico is incredibly diverse especially compared to the United States, and there are high degrees of gene flow through pollen distributions, the planting patterns, and informal systems of sharing and trading.

Most of the corn in Mexico, over 50%, is produced by very small-scale farmers, usually holding much less than eight hectares of land. Largely using animal and green manures, they are overwhelmingly managed without mechanized means of plowing and harvesting. Animal and human labor are predominant, so the burning of fossil fuels is minimal. Farmers use their own seed or trade with similar farmers. The system frequently used is that of the small holding *milpa* approach that follows the ancient way of growing by mixing corn, beans, and squash in the same fields, which is a natural method for binding soils and restoring nitrogen. They only use small amounts, if any, of commercially sold pesticides and nitrogen fertilizers. But because of successive neoliberally oriented regimes by now any previous subsidies have been all but eliminated. Peasant farmers have been forced to struggle in an ever more brutal market that has enormous consequences.

In the United States corn is produced primarily on large-scale farms, around 400 acres on the average, and is massively subsidized as we have seen—but, again, in an indirect manner in that the acreage is funded rather than the actual corn itself. From 1995 through 2003 corn American farmers received $37,400,000,000 which represented approximately 20% of their profits. Thus, somewhat hypocritically one might conclude, American trade negotiators are fond of chiding other countries for agricultural subsidies. Also, rather than saving seed or trading with fellow farmers, American farmers get seed from major producers such as Bayer/Monsanto along with their mandated contract input packages focused on herbicides containing glyphosate. The number of varieties of yellow corn available is very limited. Only 2% of the grain that American farmers sell is used directly for human consumption and the farmers themselves rarely devote any of their own land to their own families' subsistence needs. As we have seen, the American style of farming involves huge amounts of industrially derived fertilizers and pesticides as well as being completely dependent upon advanced mechanization and the use of much fossil fuel.

In Mexico households and communities are largely responsible for maintaining genetic diversity, improving crops, seed proliferation and distribution, food production, and consumption. It is all contained within localized peasant and small farmer networks of interaction. In the United States, these are all separated institutions and represent long chains of commercial, profit-oriented concentration. Culturally there are profound differences. In the United States beyond its commercial value, there are no cultural connotations attached to corn except for a very tiny proportion as found with Native Americans such as in the Southwest Pueblo peoples or among Iroquoian peoples in the Northeast for their ceremonies and preservation of "Indian corn."

In Mexico, corn is profoundly significant. Especially for Indigenous groups, both as farmers and consumers. Mexicans value highly extremely diverse varieties of foods, ceremonies, and different growing conditions. Indigenous people and those of mixed ancestry are very numerous in Mexico. Mayan and other indigenous groups venerate corn gods and goddesses with particular ceremonies, and corn goddesses are importantly associated with fertility. Mexicans often see themselves as the "Children of Maize" and look upon maize as a living mediator between land and people. Corn is often seen at the very core of creation stories. *Maize is the staple food of Mexico with much more meaning beyond simple economics or as commodities.* There were over 12.7 million metric tonnes consumed as food in 2003 with 126 kilograms consumed per Mexican. Maize occupies more acreage than any other crop, and farmer varieties had been accounting for almost 80% of that. Over 600 dishes are based on corn, and Mexicans tend to eat it in some form for breakfast, lunch, and dinner. September 29th in Mexico is celebrated as the "Day of Maize" and corn is noted for its long history in the development of Mexico and sometimes serves as a source of pride in peasant resistance to repressive national governments. In Mexico, as cannot be repeated enough, corn is essential to Mexicans culturally and economically. Its 3,000,000 producers account for more than 40% of the agricultural labor and 8% of the total labor force. Here in contrast to most of the users of corn in the world, the large portion of 68% is used directly as food. It accounts for about 59% of energy intake and 39% of protein for humans, and that largely comes in the form of tortilla flatbreads that are used in multiple ways in their cuisine.

The problem has been that, since the North American Free Trade Agreement of 1994, cheap American yellow corn has been flooding the Mexican market. By 2003, it accounted for about one-third of the domestic market and the prices for corn had fallen by about 70%, placing approximately 15,000,000 people, especially the 3,000,000 small-scale farmers, who depended on the crop, in economic peril because of declining incomes and being unable to afford basic health care and education among other things. This has forced many to seek other forms of work including at the *maquiladoras,* manufacturing zones in Northern Mexico, including ironic surges (at least 300,000 a year) in migrations to the United States—primarily males seeking jobs to support families back home, with women left to struggle to maintain the farms.

Besides the NAFTA agreement, the governments of Mexico have been devoted to neoliberal models, ignoring their large rural poor and concentrating on big business and transnational corporations. The main winners because of lower corn purchase costs in Mexico are soft drink processors, large livestock farms, and the two main tortilla producing companies, Maseca and Minsa, with direct links to Archer Daniels Midland and Cargill. As *Grain* (2018, https://grain.org/en/article/5906-mexico-the-dangers-of-industrial-corn-and-its-processed-edible-products) reports, one of the reasons Mexico is now importing 7 to 10 million tonnes of American corn annually is because of the cheap labor available and access to a huge Latin American market. There has been a large expansion of the food processing industries. The genetically modified corn from the United States is mixed in tortilla making with non-GMO Mexican corn, as well as in cereals, crackers, cooking oil, and soft drinks, as with the situation described by Pollan (2006) in the United States. Processed food production in Mexico was worth $138 billion in 2014, and Mexico is now the eighth largest processed food producer in the world, in an industry worth $4.9 trillion in 2014. The corporations producing such food in Mexico are largely transnationals such as PepsiCo, which has 17 plants with revenues of over $3.4 billion, producing brands such as *Doritos* taco chips. Tied to this has been the emergence of American-styled consumerism with, for instance, Walmex (Walmart's Mexican subsidiary) and the stunning growth of convenience stores throughout Mexico, especially Oxxo, with 16,000 stores, owned by Coca-Cola subsidiary Femsa. This all provides easy outlets for purchasing products containing high fructose corn syrup, a major health problem in the United States. Mexico, once a land noted for malnutrition issues associated with low body weights, is now the most obese in the world with 3 out of 10 considered obese and 7 out of 10 overweight. Along with all that this entails, *Grain* (2018, https://grain.org/en/article/5906-mexico-the-dangers-of-industrial-corn-and-its-processed-edible-products) points to the controversial issues of Mexican consumers being now frequently exposed to transgenic corn and glyphosate in their diets, even though the production of GMO corn is illegal in Mexico.

While all these trends were occurring, it was estimated that the subsidies to American farms exporting to Mexico were worth between $105,000,000 and $145,000,000 (out of a total of roughly $10,000,000,000 a year in taxpayer handouts for corn), with a total of subsidies from the exports to Mexico alone being equivalent to the total income of 250,000 corn farmers in the state of Chiapas. This is an extremely unlevel playing field, and they are in effect competing with American taxpayers and the largest national treasury in the world. Yet the real beneficiaries are Cargill, Bayer/Monsanto, and Archer Daniels Midlands as well as the largest of American farmers. The situation hurts Mexican farmers deeply but does not help the American rural poor. The lower costs of importing corn have not benefitted the consumer with cheaper tortilla products, and they complain of poorer tastes and nutrition (Oxfam 2003, https://oxfamilibrary.openrepository.com/bitstream/handle/10546/114471/bp50-dumping-without-borders-010803-en.pdf?sequence=1).

The biggest threats to Mexico, besides the socioeconomic disruptions, have been the loss of considerable biodiversity in corn and other fauna and flora, the emergence of industrial-style farming with the large commercial growers leading to nitrogen fertilizer and pesticide use and runoff, and unsustainable irrigation water use. The biodiversity of corn is especially significant because it is from the Mexican landraces that present and future adaptive diversity of all corn varieties depend. It helps, then, to provide the whole world with the basis of food security demonstrated by the remarkable variety of niches that have emerged out of well over 5,000 years of domestication in Mexico.

Previous to NAFTA, the Mexican government had organized agricultural policies and systems around national food security through guaranteed prices for the crop and ceilings on the price of corn products, principally tortillas, for consumers. After the flood from the U.S. maize prices for Mexican producers dropped from 807 pesos per tonne in 1994 to 559 pesos per tonne in 1999 and the ceiling on consumer prices was abandoned. Tortilla prices soared. So, both consumers and small-scale producers suffered. The small-scale *milpa* peasants lost because after meeting their family's needs there was little to be gained in selling the rest on the market to meet needs required through cash. That set the stage for the economic crisis leading to mass migrations and social unrest. With much insecurity among rural peoples. Also, the threat of contamination, as it is viewed in Mexico, by GMO corn eventually destroying the viability of traditional corn has been a major public concern with a few cases of exposure already having been documented.

In attempting to deal with these risks a network of local and transnational has organized "In Defense of Maize" to mobilize the public (Baker 2008). One national campaign in 2007 has been labeled "Sin Maiz No Hoy Pais" or "Without Corn We Have No Country." Local food networks are being organized to attempt to bring consumers and small-scale tortilla makers away from the use of industrialized corn flour, often mixed with the American corn back to the traditional methods daily used by Mexican women to maize dough whereby the maize is soaked with lime and then ground into *masa,* a wet dough, intended for the making of highly nutritious tortillas. The process was noted for its capacity to enable nutrients to be better absorbed into the human digestive system. This has become much less practiced since NAFTA and the emergence of large-scale consumer outlets such as supermarkets where, already in 2008, 49% of the consumption of tortillas was based on the industrial system of dry corn flour. Baker (2008) goes on in her article to describe how several networks of arranged traditional growing of *milpa* styled corn along with contemporary techniques from scientific-based agroecology to distributions to small tortilla-making factories, restaurants, and alternative consumer outlets. The movements though do face uphill battles though with Mexican consumers rapidly taking on globalized consuming and eating habits favored and induced by transnational corporate interests.

One would have to conclude that corn-growing as a mainstay of Mexican society is at a long-term risk. While such massive transitions have been seen before as with Europe, Canada, and the United States and the "excess" of rural populations have been absorbed into urban and manufacturing occupations, that is not likely to occur smoothly or peacefully in the Mexico case because of the actual increases in poverty, health problems associated with new consumption patterns, and the conflicts associated with the Mexico-U.S. borderlands that offer no easy solutions.

Soy

Parallel to corn, Raj Patel (2007) provides an introduction to soy as another exemplar of the dangerous consequences of neoliberal destructive influence in contemporary global agriculture. This case study is an apt illustration of the major themes of this book—power, ecology, social disruption, the corporation, and the role of the nation-state. Besides being a direct animal and human food, soy has many other uses in industrial systems and in food processing. It contains an important ingredient—lecithin—that allows food processors to mix fats and water. According to Patel about three quarters of the foods on supermarket shelves contain soy and its derivatives, and it is widely used in fast food. It is a key ingredient in margarines and vegetable oil and has been added to biodiesel fuels, inks, and it has even been used as lubricants, and for seating fabrics and wiring in automobiles. Carpets have been made from it and it has been combined with recycled paper to manufacture products that have the fibrous consistency of wood.

It can be an important source of food for humans when processed into tofu because of its high concentration of amino acids as a source for protein building as well as with other products for people on restricted diets such as soy milk, soy flour, infant formula, soy sauce, and soy yogurt. Yet these all require much processing. Still, as with corn, the vast bulk of its production goes into animal feed for chickens, hogs, and cattle—80% of it. The rest goes into processed food, manufactured products and oils, and much capital investment is required.

Soy grows best under wet conditions with loamy, non-sandy soils, and with long, warm growing seasons that are found in the American Midwest and some parts of the Great Plains. Illinois is the epicenter, and the major transnational corporation dealing in all stages of its production, as with corn, is Archer-Daniels Midland located at Decatur, Illinois—and most notably beyond in the processing dimensions of the industry. Others include Cargill of Minneapolis in the shipping and trade and Bayer/Monsanto of St. Louis since the overwhelming amount of soy, as with American corn, is now genetically modified bolstered by the pesticide Round-Up to which the seed has been made resistant.

During World War II the industrial dimensions were further developed, including the manufacture of glues. The United States became *the* major producer, outstripping Manchuria. After the war, the United States used its enormous power and wealth to heavily subsidize farmers for massively increasing production and used its international muscle to manipulate trade agreements so that it held onto 90% of the world's export market until the end of the 1960s.

Then, though, prices started to fall, and the United States began relinquishing its virtual monopoly to Brazil. Brazil contained much poverty, corruption, disruption, and violence. This developing country was ruled by oligarchic landlords who kept the majority of the population, 62%, landless and in debt service to them. Inflation was rampant as were food riots. Unions and peasants tried to organize and there was much civil unrest. Right-wing dictatorships took over after 1964, frequently bolstered by military coups, and sought means to improve and stabilize the economy and reduce rural discontent. They looked for commodities that could improve Brazil's balance of payments and its growing international debts. While the soy industry had had a small presence in Brazil, the early 1970s saw several events opening a major opportunity for Brazil. Overfishing of anchovies, and, a major collapse of African peanut production led to more reliance on soy, but American surpluses of soy had been depleted in response.

Brazil captured 18% of the global market by 1979. Much of the financing came from Japanese interests, and the government also developed policies for land expansions and provided no-interest loans for the processing plants. Major American transnational organizations also struck claims—most notably Archer Daniels Midlands, Cargill, and Monsanto, and expansion continued not only in Brazil but also Argentina, Paraguay, and Bolivia. Yet another global crisis occurred with the virulent outbreak of mad cow disease in Europe in the 1980s, which led to the preventative slaughter of millions of cattle increasing the demand for only plant sources of animal feed containing high amounts of protein.

In Brazil, the production increased by 77% between 1995 and 2004—much of it in the Matto Grosso region of West Central Brazil that contains much flat land, forests, and savannah-like mixed forest and grasslands. Many thousands of acres of forest were cut down to accommodate the production. The *cerrados* or savannahs were at first particularly the target for both expansion in cattle ranching and very large agricultural soy plantations. Much, however, went into the actual tropical forests turning them into *cerrados*. Land has been very cheap, and the production costs are much less than in the American Midwest.

As Patel (2007) points out, the ecosystem there is extremely important and fragile. The *cerrados* sit on top of one of the world's largest aquifers; consequently, the irrigation system required for soy is constantly draining the system. The *cerrados* are extending into the Amazonian Rainforest itself, led first by pioneering ranchers and then soy planters. The reality of ever-increasing damage is continued, especially once roads are established leading to all sorts of in-migration. The Mehinaku people, indigenous to the area and already made vulnerable because of infectious disease with numbers barely over 200, found that development has severely undercut their security of maize and manioc farming along with subsistence hunting and fishing that require access to extensive territories. For people of color and the landless, wages are low and jobs are not as available as they are in ranching or sugar cane. Lawlessness abounds, and Patel (2007) tells us that the International Labour Organization estimates that there are even circumstances of slavery in the regions ranging from 25,000 to 40,000 persons held in captivity. Brazil has attempted to stop it, but even government inspectors have been murdered. Brazil has also introduced laws in the 2000s to stem the scope of illegal forest clearings to create *cerrados* principally for ranching and soy plantations. Yet because of widespread corruption and oligarchic governments, rich landlords continue to get away with their crimes.

Patel (2007) points out that there have been threats of trade wars and verbal taunts between the United States and Brazil regarding soy farming. Americans correctly tend to see the industry in Brazil as being highly destructive of the Amazon, the world's largest carbon sink. Brazilian oligarchs counter, also correctly, that the American government provides unfair subsidies both to companies and to farmers growing the crops.

It is also true that American corporations are big players in the Brazilian context. Cargill serves as the primary exporter of both American and Brazilian soy. It has over 8,000 employees in 191 municipalities in 17 states, 19 processing plants, 182 warehouses, and 5 port terminals (Cargill Annual Report 2014, https://archive.org/stream/5687727-Cargill-Annual-Report-2014/5687727-Cargill-Annual-Report-2014_djvu.txt). It exports much soy especially to European suppliers of McDonalds—feed for chicken and beef. As fast-food chains expand overseas there has been more pressure for quality crops all along the food chain. Cargill has been seen as a major promoter of land clearing. Bayer/Monsanto Corporation the major promoter and developer of GMO crops globally has also been a big player. There has been controversy almost always surrounding this corporation. GMOs were illegal until 2005 when a regulatory system was put in place that permitted it. Previously that did not stop illegal amounts of Round-up ready soy being introduced to the soy-growing regions. Today approximately 90% of soy crops in Brazil are GMO with Monsanto having the largest share of the market. However, glyphosate-based pesticides—the essential tool accompanying Monsanto's products—proved to be somewhat less efficient than expected in that significant amounts Round-up resistant weeds thrived. Farmers in the state of Matto Grasso were in the midst of a major suit with the company over royalty issues (Reuters 2018, https://www.reuters.com/article/us-usa-pesticides-soybeans-insight-idUSKBN1FD0G2). Beyond that, large transnational corporations such as Cargill, Bunge, and Archer Daniels Midland finance over 60% of soy produced in Brazil and they own three quarters of the facilities devoted to processing and transportation to Europe.

Transportation routes have been extended with highways through the *cerrado* zones and eastward to port facilities. These routes open up regions for settlement but the agriculture for human consumption offers little adequate nutrition. These truck routes are dotted with broken down socially dysfunctional towns that have cheap taverns, prostitution, strip bars, and high crime rates and violence, especially against women. The large firms have operated in alliance with each other and have been responsible for much of this decay. For instance, Patel mentions the state of Parana with its main soy exporting port having a law that prohibited GMO soy and its transportation. The corporations conspired and had their soy sent to another port in another state. With regard to lobbying, Cargill tends to form alliances with Monsanto, ConAgra, and Novartis with Archer Daniels Midland and all sometimes work together to send out a unified market signal (Patel 2007).

Patel (2007) claims that it is in both the U.S.A. and Brazil that the small farmers who benefit the least and are hurt the most. In Brazil, it is clear that indigenous peoples and landless laborers are severely damaged. If one were to look alone at aggregated data, it would superficially appear that both countries have benefitted. But the rewards are unequal going to large farmers, especially in Brazil, the processors, and corporations. Declining standards of nutrition and food consumption documented by the U.N. Food and Agricultural Organization are reported for large parts of Brazil, especially in the *cerrado* regions. Large farms dominate—in 1996 the 1% in Brazil, those over 1000 hectares dominated covering 45% of the arable land. Many families have been driven off the land— over 5,000,000 families with about 150,000 camped on road allowances. The upper 10% of the population had an income that was nineteen times greater than the lowest 40%. There has been an ongoing trend whereby large landholdings owned by either families or large corporations expel squatters and force small landowners to sell out as the heavy mechanization of agriculture to meet global market conditions. A good percentage of the rural labor force has been made superfluous. Many of these people have been forced to migrate to the slums of the large cities or in government programs to take them further into forested interiors to cut down yet more rainforest for development with many of those programs leading to severe soil erosion.

Soy is not confined to the United States and Brazil. Its production is expanding rapidly into the "Cone" of South America. Correia (2017) begins his article about Paraguay by referring to an advertising campaign by the bio-tech seed giant Syngenta that referred to the "United Republic of Soy" as an imaginary state involving Paraguay, southern Brazil, Argentina, Bolivia, and Uruguay. The contention is not far off the mark. Its presence has played *the* major role in reshaping South America ecologically and economically with huge political and social consequences. A Spanish term *sojización* now used widely refers to that. Its meaning refers to the "territorialization of state-space by the soy industry and soybean plants" (Correia 2017: 2). There have been huge and violent dispossessions of indigenous peoples and peasant farmers and the types of food grown are now very limited, which is a major threat to food sovereignty.

Conclusions

Grains such as wheat, rice, and barley along with corn are the staple foods that feed over eight billion humans. Yet very large portions of those crops, most especially corn and soy, are being devoted to animal feed for the protein needs of growing numbers of middle-class people. Scientific and technologically oriented crop development has provided seemingly miraculous increases in yields that may have reduced the threat and reality of famines. But such yields are also responsible for the large-scale global population increases that we have been experiencing from the second half of the 20th century onward. And as we have seen from the discussion of the Green Revolution, much of that growth in crop yield is due to the influence of oil interests via the Rockefeller and Ford Foundations over American foreign policy. Taking into account nitrogen fertilizers, the development of chemical pesticides, and mechanized labor in farming, we could say that we are "eating oil"—hydrocarbon energy converted into carbohydrate energy at the ratio of 10 to 1. Hybrid and now genetically modified plant breeding, placing the innovation of seed varieties in the hands of laboratory, and experimental farm scientists rather than farmers themselves have played a very large role in creating these shifts in yields. Yet as should be clear, scientific, and technological these innovations are not neutral or impact-free. The consequences of these revolutionary technological changes are many and profound.

We have seen in this chapter how two plants, corn and soy, have acted as imperial crops in the novel sense that Michael Pollan (2006) suggests—plants also have played a co-evolutionary role in domesticating us by extending their domains through humans becoming dependent upon them. Within the human realm, the American, Mexican, and Brazilian corporate business and government officials have seen to soy and corn expansions benefitting their self-interests and neoliberalism. Cheap highly subsidized corn with large-scale industrial production, marketing, transportation, and processing systems have come to dominate American agriculture in very many respects. This has significant health and environmental consequences as documented here. Then the imperialistic dimensions, as even implied by the phrase "King Corn" are greatly augmented by American corn being dumped in Mexico. With soy, its similar imperialistic expansions are found in its direct export to other countries but also in the export of industrial means of agricultural production along with the presence of gigantic transnational corporations to huge swaths of South America. In these Latin American countries, the production of soy or the import of corn has led to huge and painful displacements of people. In the case of Mexico, because of massively economic uncertainties, large numbers of people are leaving corn growing and attempting to migrate to the United States or to the crowded industrial parks in Northern Mexico. In South America, indigenous peoples as well as peasant settlers are often violently pushed aside to make way for huge mechanized soy plantations. Many are forced to migrate and live in poverty in city slums or become landless agricultural laborers. In both cases, there are environmental losses of water tables, biodiversity declines, deforestation, and pesticide and fertilizer contamination. In all the cases, including the United States, there are the consequences coming from the dominating power of transnational corporations.

Such examples should serve as warnings to farmers in the global South as neoliberal interests seek to expand such systems there.

4 Livestock Production

Since dinosaurs became extinct and mammals dominated the Earth, the basic pattern has been for an abundance of herbivores grazing on grasslands, essentially wild grains, with a smaller number of carnivores feeding upon them. We evolved as hunters in similar relationships. Through our massive interventions, we now completely dominate as carnivores with our domesticated cattle, sheep, goats, hogs, and poultry. A significant part of the Earth's biomass has been transformed through our interventions with the total weight of ourselves combined with our domesticated animals overwhelming that of the remaining wild mammals (Lewis and Maslin 2015). Besides the questions of whether or not there are too many of us and if we eat too much meat, its production has become thoroughly industrialized and dangerously threatening to environments.

As with most contemporary forms of agriculture in the developed world and increasingly in the developing world, livestock production is now vertically integrated, dominated by a small number of powerful corporations. Five companies are responsible for the bulk of meat production in the United States (Mighty Earth (2017)—with Tyson Foods and Smithfield dominating along with feed suppliers Archer Daniels Midlands, Bunge, and Cargill. Tyson controls 20% of the chicken, beef, and pork markets, so is the biggest player. Twenty-four billion pounds of beef, along with 40 billion pounds of chicken, and 25 billion of pork were produced in the United States in 2015. ETC Group (2017) shows the rise of the Brazilian JBS transnational to the number one global position.

The Virtual Extinction of the American Buffalo and the Rise of the Beef Industry

This example is a classic case study in political ecology. I am referring to the virtual extinction of almost 30,000,000 North American buffalo on the North American Prairies in less than 50 years. Euro-American and Euro-Canadian settlers moved onto the short and long-grass prairies regions of the United States and Canada bringing with them the grain and cattle industries and, in the process, completely changing both the fundamental ecologies and human relations to them. The transformations saw the mass commodification of meat, the rise of corporate-controlled feed grains, and the ultimate domination of agribusiness

DOI: 10.4324/9781003495673-5

through meat-packing corporations controlling the primary producers of ranchers and farmers. Thousands of indigenous people were massacred or marginalized on reserves while serious issues of environmental degradation continue at alarming rates.

Wes Jackson (2010) sees this as a continuing ecological crisis for which we have little time left to adjust to further disaster that could completely wipe out agriculture in the region. Native Americans with their small-scale maize agriculture and buffalo hunting, Jackson contends, provided much better stewardship for the North American Prairies. In his opinion, there are very few years left for any kind of sustainable agriculture for this region be it grain farming or beef production. The history and consequences of this ecological and social disruption are well told in Jeremy Rifkin's (1992) *Beyond Beef: The Rise and Fall of the Cattle Culture,* neatly summarized in anthropologist Richard Robbins's (2011) *Global Problems and the Culture of Capitalism* and Joshua Specht's (2019) *Red Meat Republic*—all three of which I will rely on here.

Robbins (2011) begins by discussing the large-scale environmental inefficiency and costs of the contemporary beef industry. Eighty percent of grain that is produced in the United States goes to feeding livestock, plus two-thirds of U.S. grain exports are used in feeding animals in other countries. Half the water consumed in the United States is used in the growing of feed for cattle with the amount of water used to produce ten pounds of beef being equivalent to that used by the average American family in a year. It requires 15 times more water to produce one pound of beef protein as it would to produce one pound of plant protein. There is much pollution generated by massive feedlot operations—with feedlot steers producing as much as 47 pounds of manure a day, and the methane released by them being a major contributor to global warming since methane molecules have 20 times the capacity to capture heat as does carbon dioxide. There is further greenhouse release from the slaughter, butchering, packing, refrigeration, transportation, and the cooking of beef. Much rangeland has become degraded by overgrazing. Western public lands of 306,000,000 acres are devoted to rangeland, and its overuse threatens. Raising beef is an inefficient use of converting plant food—at the time of Rifkin's (1992) writing the conversion was 157 million metric tonnes of cereal and vegetable protein converted into 28 million tonnes of animal protein. Beef eating has been associated with rising levels of cancer, obesity, and cardiovascular disease. Americans created a massive beef-eating consumer culture and are exporting this food culture—especially its fast-food, burger system to the world.

Meat is used primarily as just a supplement in most societies; otherwise, diets center on complex carbohydrates—rice, pasta, tortillas, and bread. It took a number of significant steps involving both political, economic, and political ecological transformations, including the capitalist manipulation of consumer tastes, to change that.

As part of an Old-World complex emerging in the Caucasus region much further east, the central plateau region of Spain had been ideal for cattle raising and through extensive and large open-range grazing system derivative of the Roman *latifundium* or plantation, hierarchically organized with many subordinate workers operating on horseback. Settlers in the Spanish Americas established the ranching system in what is now Mexico, Central America, Texas, Uruguay, and Argentina, and large numbers of cattle were imported from the Old World. The ranching-cowboy cultural system was consolidated and further elaborated through its spread northward. Meanwhile in the British colonies, at first and as a continuity of English tastes and adaptation, hogs were preferred in the densely forested regions of the Eastern Woodlands where they could be kept in large numbers, and the pork preserved and flavored through salting, pickling, and smoking.

By the late 1700s a shift started in Britain, the world's leading superpower. The prosperity there led many to seek a meat-dominated diet. First among nobles and landed gentry, there emerged a prestige consumer taste represented by exclusive "beef eating clubs" and attempts to meet the rising demand by English landlords were tried in Ireland after the Potato Famine. These did not work, so overseas sources were sought and refrigeration was developed for steamships in the 19th century. Those trends assisted the development of a lucrative British market. British investors dominated the early beef industry of Argentina. Meanwhile, beef was spreading as a newly preferred food among the British middle and working classes—the soldiers and sailors of the Empire were given rations of ¾ of a pound a day. Britain became an industrial powerhouse, its population expanded greatly, and its working population became more prosperous with raising consumer demands for sugar, tea, and beef. Canned corn beef became a popular staple among the English working class and what has become an English tradition was the Sunday, roast beef family dinner, complete with Yorkshire pudding and various condiments to enhance the flavor.

By the late 1840s, investors had started to consider a yet untapped source— the American Southwest, principally Texas that had been acquired by force from Mexico during the Mexican-American War of 1846–1848. American settlers in Texas found vast herds of wild longhorn cattle free for the taking and adapted to the Great Plains that eventually stretched even further to the North. American settlers started to round them up or steal them from Mexican ranches in vast numbers. A number of problems, though, stood in the way. Driving them overland to the American Midwest was too costly. There were still huge herds of buffalo—in the many millions—in the region and much further to the North that would be in competition for open grazing in the huge grassland regions. And there were the thousands of Native American peoples who depended upon the buffalo and were considered by the Americans as hostile and blocking their claims of Manifest Destiny.

Constructing railways was one approach that led to the Euro-American ascendancy on the Plains. They had been extended into Kansas and Nebraska that served as rallying points for cattle drives from Texas. The sliding pen had been invented that allowed the cattle to be driven into railway cars to be shipped to places such as Illinois or the Eastern Seaboard. One of the policies, explicitly supported by cattlemen, investors, railroads, and the U.S. Army, was to exterminate the buffalo to make room for unfettered cattle range expansions. This was done in various ways—as the railways expanded westward, passengers were issued with rifles to shoot them for sport from their moving cars and hunting as a tourist pastime was encouraged. Carcases were left to rot on the prairies. Professional buffalo hunters including William Cody and Buffalo Bill killed hundreds of thousands with the pelts used for coats and the bones for fertilizer.

The period 1870–1880 was a time of wars waged on Native Americans by the U.S. Army under General Phillip Sheridan. He openly approved the slaughter of buffalo and viewed the replacement of their range by cattle ranching as superior because it was the best way of defeating and pacifying Native Americans. Eventually, in all of the United States, only several hundred wild buffalo remained and Native Americans were confined to reservations. Then, ironically, ranchers contracted by the government to feed starving Indians became rich by overpricing and serving rancid meat, and were frequently able to graze their cattle on reservation land for next to nothing.

The next concern for the British market was that western cattle were too lean for taste. A solution was devised whereby they were transported by rail to the American Midwest, fattened, and then sent to ports and steamers headed to Britain. For that, there was a large amount of investment by British banks and other companies in very large ranches in the Northern Plains states and territories of Colorado, Wyoming, North and South Dakota, and Montana. By the 1870s, breeds other than Texas longhorns more suitable to market tastes were added to American herds. Before the end of the 19th century 90% of the beef imported in Britain came from the United States. Yet into the 20th century there was a reaction against such dominant foreign ownership and Americans took over. Many of the British mega-ranches had suffered bankruptcy during the devastating winter storms of the 1890s.

Due to Midwestern farm interests in selling corn and fattening cattle at feedlots, the United States Department of Agriculture initiated a system for grading beef—with the highest prime Grade A going to the fattest cuts. That encouraged a system feeding corn to cattle and limiting their movements—both factors detrimental to their health in that grass-fed ones are healthier regarding ease of digestion and cattle are intended to wander and graze. Such a system also signaled to the consumer that the best cuts were the fattest.

Eventually federal and state policies favored the centralization of butchering, rendering, and packing, and Chicago dominated as a railhead center both for the marketing and transportation of grains as well as that of livestock and cattle's further transportation to the East and Britain. Refrigeration became more sophisticated and contributed to the value-added dimensions of the beef industry. Five companies, dominated by Swift and Oscar Meyer, controlled the meat packing industry that had become industrialized through the killing, butchering, and packing "disassembly" line where workers at stationary positions further butchered slabs of meat that were brought to them. As Specht notes (2019) the packing companies manipulated the market so that they became the most dominant players surpassing the local ranchers and even the railroads. In the early part of the 20th century, federal laws allowed for ranchers to appropriate huge amounts of public land for grazing.

A big milestone was the invention of the hamburger. The hamburger is associated with Americans on the move and the infrastructures associated with new eras of mass prosperity. Supposedly, this unique American dietary staple was invented when a restaurateur ran out of pork sausages served on buns at an Ohio State Fair in 1892. He substituted ground beef and that minor innovation became popular at the Saint Louis World's Fair two years later. A full-fledged hamburger chain was established in Kansas City in 1921. Then came the huge advances in the automobile culture. After World War II, the U.S. government expanded its highway system to include 41,000 miles of four-lane superhighways at a time of unprecedented mass prosperity and the purchase of large numbers of personal vehicles with extremely low gas prices. This mobility along with feeder roads leading to the downtowns of major cities meant easy access for the rapidly constructed suburban bedroom communities. These trends along with rapidly rising prosperous middle classes, cheap mortgages, and mass prefabricated housing construction, all led to the establishment of the suburbs as the dominant American settlement pattern. Feeder highways containing strip malls, service stations, and fast-food outlets became standard. The pioneer occupant of this monotonously repeated landscape was the McDonald hamburger chain, and many other imitators soon followed because of the immense profits in serving people rushing to and from work looking for fast meals that they did not have to prepare themselves.

Hamburgers became significant for at-home diets as well. Hamburgers have an advantage over pork in that they do not require excessive cooking to be disease-free. Busy housewives as they became absorbed into the workforce through the 1950s and 1960s, started to make use of fast-food ingredients for their families, and frozen hamburger patties were especially convenient along with frozen french-fried potatoes. Added to that was the widespread American custom of the outdoor barbecue that made use of the hamburger in a context of social entertainment and hospitality with friends and neighbors.

The United States Department of Agriculture again came to the support of an integrated beef industry in 1946 with its definition of the hamburger supporting both the corn and beef industries. A hamburger can contain up to 30% fat but that fat must be from animals. This allowed the meat packing industry to use the scraps of fat left over from butchering—a cheap source of binder. It also reinforced the productive capacity and profits of corn growers as well as feedlot owners in the American Midwest. During the postwar period, the beef and hamburger culture became central to American society. Yet serious health concerns associated with cancer and cardiovascular disease have emerged so there has been some reduction of its consumption in the United States. Yet as a commodity it is deeply entrenched and has significant economic, political, and social meaning that make it difficult to reform.

The United States has been a major producer of beef because of the history that we have outlined here, but the cultural complex associated with its production and consumption has spread globally and has severe implications. The United States now produces about 9% of the world's beef but still consumes about 28%. That difference is made up from new sources of import from Latin American countries—both Central and South America. That has been occurring since the 1960s, and the U.S. Department of Agriculture has been involved in arrangements with countries for establishing USDA criteria of inspection, labeling, and other procedures. Beef from these areas can be produced at 40% less cost than in the United States. The World Bank has provided preferential loans for establishing beef industries in these countries. Cattlemen organizations have been initiated there and they lobby fiercely for expansion often using violent tactics against indigenous people and settlers and generate huge negative impacts on tropical forest sustainability. Peasants and indigenous peoples are evicted in large numbers and forced to migrate to already overcrowded cities exacerbating poverty and unemployment while providing surplus amounts of very cheap labor for neoliberal manufacturing interests. Ranching in these regions can be done profitably on a very large scale and its expansion in these regions only leads to further concentrations of wealth.

Beyond these production factors, the hamburger culture is becoming global. McDonald's has almost 3,000 outlets in Japan and plans or has built about 5,000 in China with significant numbers in the Philippines, Korea, India, Taiwan, and the Near East. (https://www.statista.com/statistics/256049/mcdonalds-restaurants-in-the-emea-region/). Burgers are often associated with modernity and a higher status where they are a novelty. Americans have reduced their per capita intakes, but the market is more than made up by this expansion globally.

Again, this case study is an exemplar of the political ecology approach. Cultural history shows the origins of a major production and consumption trend and how it was all initiated by powerful capitalist interests as well as those of the nation-state. Here we can include ranchers, railroads, bankers, feedlot operators, feed growers, and ultimately agribusiness interests in the corn trade and especially meat packers, the World Bank, national governments in Latin America and of course, the post-Civil War U.S. nation-state in its "pacification efforts" on its western frontier. There has been much social dislocation associated with it—the genocide and displacement of large numbers of Native Americans with similar contemporary impacts in the tropical regions of Mexico, Central America, and Brazil.

Then there is the astonishing short period leading to the nearly complete extinction of the buffalo. The buffalo had been virtually perfect in its superior adaptation to the vast Great Plains. Buffalo was climatically adapted to extremes of cold surviving the worst winter storms; they easily dug deep through snow, reinforced and thrived on perennial grasses with deep roots in deep black soils that preserved moisture in a very dry environment. Native Americans were also extremely well adapted to this animal making use of all body parts in their material culture and by no means threatening its sustainability.

Not only was this near extinction caused by human action but, also, we have had major negative transformations on a huge landscape devoted to seed and grain crops. Finally, we have to be concerned with the huge contributions of this industry to global warming.

Cattle Ranching and the Brazilian Amazon

With reference to meat production, the country with the most cattle for beef (and leather) production is Brazil with about 232,000,000 head in 2019 (https://www.ers.usda.gov/amber-waves/2019/july/brazil-once-again-becomes-the-world-s-largest-beef-exporter/). While supplying a large domestic market in a beef-eating culture, Brazil is the major global exporter of beef. Yet the ecological costs for this ascendancy may ultimately be much more severe than even the near extinction of 30,000,000 North American buffalo. That relates to the massive deforestation of the Amazon tropical forest with huge ecological chain events and socioeconomic impacts including the loss of the world's most significant carbon sink, major destructions of biodiversity, hydrological and weather disruptions, massive dislocations of poor farmers and indigenous peoples, and much violence and corruption. Land grabbing and even bonded-labor, verging on slavery, are part of the ugly social heritage (Hecht and Cockburn 2010). Forty-eight percent of all global tropical forest loss can be found in Brazil and three quarters of that can be blamed on ranching. Brazil has been ranked as the third largest greenhouse gas emitter after China and the United States and 50% of that is due to continued forest loss.

The Massachusetts Institute of Technology's Amazonia Project (MIT 2006) on Brazilian ranching shows how it severely exacerbates biodiversity loss through cattle eating grass right down to the dirt level, removing virtually all nutrients not to be replaced and with the habitat unable to restore itself. The soil is compacted through grazing, and plants with larger root systems that preserve water can no longer penetrate the ground. During the rainy season, nutrients are leached out beyond the roots due to the removal of protective canopies composed of larger broad-leaved plants, and during the dry season, plants are sun-drenched and limited to drought-resistant species. For every quarter pounder of hamburger consumed in the United States originating in Brazil, 55 square feet of Brazilian tropical forest is lost, and in spite of the claims by many fast-food corporations that they do not use rainforest beef, that cannot be confirmed because such meat can easily be processed as domestic.

Cattle ranching and beef exporting on such scales are relatively new to Brazil, limited until the 1990s when the spread of mad cow disease outside of the country and currency devaluation led to the rapid growth of the industry. There is now much complexity in its value chains—small and large ranches provide cattle to fattening farms and then to slaughterhouses and meat packers. Tens of millions of hectares are ranched within hundreds of thousands of ranches, but slaughtering and packing are in the hands of four companies including JBS, the Brazilian company now the largest in the world. An attempt to restrict that devastation occurred on October 5, 2009, when the four major meat packers signed an agreement with Greenpeace to only purchase meat from ranchers who had legally established their land holding before that date. As well, other agencies due to international pressure have developed policies to stem deforestation (Walker et al. 2013).

In spite of international ethical pressures on the market and internal policy accommodations that have attempted to stem the rapid deforestation, much of it continues legally and illegally. Anthropologist Jeffrey Hoelle (2011, 2014) examines why this is so with a detailed study in the state of Acre. The more recent increases in cattle expansion there had been the largest of all the Brazilian states. Ironically, they are largely due to its adoption by small-scale agricultural settlers and rubber tappers who had formerly been vigorously opposed to ranching. This is due to a changing political ecology as well as some significant cultural shifts. Before ranching made its way into the region in the 1970s, traditional collectors had relied on the tapping of rubber and the foraging of Brazil nuts in relatively untouched forest conditions, and at the same time relatively recent pioneering colonists engaged in small-scale slash and burn agriculture. The farmers migrated from other more impoverished and crowded regions of Brazil with being supported by government agencies, programs, and subsidies. Similarly, large-scale ranching had been explicitly encouraged through large financial incentives, and such policies had made it feasible to buy out large rubber plantations, along with other forms of cheap land and subsidies to the benefit of ranchers.

Major clashes used to occur, most especially with ranchers and the long-time rubber trappers when government policies definitely favored ranchers. However, eventually relative peace was brought to the area with the establishment of a large forest reserve for traditional foraging for subsistence and market purposes.

After 1990, with the introduction of neoliberal policies, subsidies for rubber and supports that maintained small family agriculturalists were canceled. This has placed tappers and small family farms in peril and, in the case of the tappers, the Brazil nut then became the preferred source of income. However, the prices for the nuts and the commodities produced by the farmers remain highly volatile.

For both of these more marginal groups, the consistent high price for cattle, ironically, has made the option of raising cattle, which they previously vigorously opposed, quite attractive, and many are now taking up cattle raising. They find that the money that they make from this is the only way to meet expenses and save the farm due to debt. Many have taken up the subculture with cowboy dress, Brazilian western music, and rodeos. The lifestyle is quite demanding and best suited for young men. They are able to gain jobs as working cowboys on large ranches and use the experience to persuade their families to convert to beef production. Ranching has become the preferred norm and the other forms of economy are seen as in decline, decadent, and the sources of poverty. Ranching has gained greatly in prestige, seen as romantic, and a source of wealth.

The governments, federal and state, continue to support cattle raising through extension services, the introduction of new, better, and more affordable breeds, and have been able to stop the spread of various diseases such as hoof and mouth related. Other aspects of an infrastructure are supporting the shift—agricultural supply stores, butcher shops, slaughterhouses, and intermediary services to pick up and deliver cattle to various points of development from calving to fattening ranches to slaughterhouses and to market. This all tragically supports the continuation of deforestation.

Pork Production

There has been a worrisome set of transformations in the production of hogs since the 1970s with many social and environmental impacts (Thu 2010; Thu and Durrenberger 1998; Ervin et al. ed. 2003). Production has shifted from independent, small, mixed grain and livestock operations to massive confined animal feeding or factory operations (CAFOs). Tied to this has been the emergence of vertically integrated, extremely powerful, American-owned transnational corporations such as Smithfield's and Tyson. They have established production methods focused on economies of scale that are able to provide cheap pork to consumers. One motive for this has been the rapidly expanding Asian markets with their newly rising middle classes and where pork had always been the preferred form of animal protein.

CAFOs consist of long, low, windowless, metal buildings where as many as 5,000 hogs can be tightly penned up on concrete floors with chutes and conveyor belts leading to feeding troughs. Below are drains that create flows of urine and excrement to metal channels and take them to lined lagoons where they may be treated eventually to serve as fertilizer although of a kind that is often overlooked because of toxic concentrations, especially of nitrates. While most hogs are approximately the same size as humans, they defecate about three times as much daily. This then creates a huge amount of waste on these farms that often contain multiple buildings where the hogs are kept. These effluents often make their way into groundwater in spite of protective linings and they are subject to flooding that can overrun streams and much larger bodies of water such as ocean bays where they may have severely negative effects on fisheries. Inside, ventilator fans must be constantly working because the stench is intolerable at the source and extends many miles beyond the building depending on variable winds and other atmospheric conditions.

These odors make it very hard on the workers who are already stressed in many other ways. As hidden videos have shown, they sometimes resort to inhumane treatment of hogs such as in castration or the brutal treatment of sick pigs. Although the establishment of these hog farms has often been touted as a source of off-farm, local employment, the employees often come from migrant labor sources, which may in turn lead to local conflict. Thu (2010) has been able to show that occupational health risks are quite high in this line of employment. Furthermore, there has been a tendency to situate these facilities in poor communities, especially in the rural American South because of the environmental and social controversies surrounding them.

Practically all of the industry is now under the ultimate control of only a few firms with vertical integration and long and complicated global supply chains. They either run the hog barn units themselves as company operations or they franchise out to independent producers with strict contractual conditions. Some operations will run the full gamut—from farrow to finish, where sows are forced to have three or more litters a year through artificial insemination in quite confined spaces. Sometimes there will be separate operations of farrowing to feeder pigs. It is all extremely regimented and industrialized. The firms have substantial capital flows to purchase large amounts of cheap, surplus grains such as barley, soy meal, and especially corn that are processed into pellets for feed. There is a very strict feeding schedule for the hogs at particular time of their developmental cycle so ideally their weights and sizes are about the same. Many of the eventual slaughterhouses are highly automated and process live animals very similar in size before slaughtering and rendering to make it easier for the calibration of the process. A frequent sight at the CAFO complexes is that of large trucks coming and going hauling large numbers of pigs who have not really seen the full light of day during their lives away to the slaughterhouse or to the CAFO operations to begin a new cycle of feeder pigs to finish. Then there have been wholesale removals for disposals of large piles of dead hogs that have been the victims of epidemics.

Contrast this with the standard ways hogs were raised in the past. North America, in the early 1970s, had hundreds of thousands, small to medium sized, independent operations. Some of them were specialized such as farrow to feeder or feeder to finish and some just managed hogs buying feed on the market. A very large number though were mixed grain-livestock operations owned as family farms. Corn and hog types were abundant in the American Midwest and in

Canada barley and hogs. Hogs by nature root for their food underground and wallow in mud to keep cool but to also divest themselves of pests. They are also highly intelligent and sociable. Although penned up in barns for winter protection, during other seasons they were allowed to wander and exercise outside within fenced areas with ponds for mud and water with opportunities for piglets to socialize with the supervision of sows.

Farmers improvised their strategies around growing more grain for the market or alternatively their lower priced grain stocks to feed more pigs when the latter gained a higher rate on the market. In some Canadian provinces, there were "single desk marketing" procedures through which government-regulated agencies would negotiate in national and global markets, gaining the best deals possible for the bulk sales of hog carcasses, thus serving their producer members. Such systems no longer exist, and hog producers have to operate individually on the market, switching the advantage to huge corporate buyers and sellers. During the previous period, independent producers could almost be considered a sub-cultural type of farming running in families for many generations. Although obviously not perfect, two things were important: (1) hogs were raised a lot more humanely than today, and (2) the independent operations of farmers in the hog business were well integrated into local rural communities.

The dramatic shift began in the mid-1970s, with the numbers in the United States from 1974 to 1996 reduced from 750,000 to 157,000. The years 1994–1996 were especially difficult, with one in four operations disappearing at that time. Then 3% of the largest producers produced 51% of the hogs. More of the remainder went into the franchise mode of production for the large corporations. This was a response to rising market demands spurred for a while by the notion of pork as a healthy alternative lean "white meat," and the ever-growing Asian demand. There were other factors involved, such as, as Thu (2010) points out, the flooding of the market with cheaper than normal cuts of pork in 1994 at 57 cents a pound (the lowest price since the Depression) by the largest producers, which eliminated much of the smaller competition. It all relates to economies of scale since there are benefits of buying huge amounts of feed, especially at the most opportune times, and saving them for later. The smaller producer is faced with the cost-squeeze pressures that involve factors of machinery, chemicals, seed, insurance, and debts. The advantages then go to large operations and corporations.

This transformation began in the American Midwest within the corn and soy belt. Then it expanded into the American South, far West (notably Colorado), then virtually every Canadian province, and next Latin America, especially with Brazil and Mexico being the most current centers for expansion. In my own home province of Saskatchewan there were 12,246 farms raising pigs in the early 1970s; by 2006 there were only 930. The numbers produced, though, went from 500,000 to about 1.4 million. In 1976 85% of the farms produced less than

2,600 pigs a year; in 2006 over 90% of the farms produced that amount or more (Encyclopedia of Saskatchewan n.d.).

With these types of massive corporate intrusions, it is the local community that pays the price. At first jobs may be offered but, in many cases, they turn out to be few in number, highly stressful, and eventually taken up by migratory labor that in turn suffers considerably. Respiratory and workplace injuries abound. Local residents, not even involved or benefitting from the industry frequently, suffer from the horrendous smells emitting from the barns and waste lagoons exacerbating or creating anew, severe respiratory problems. Residues contain high concentrations of phosphorus, ammonia, hydrogen sulfide, nitrogen, and over 160 volatile organic compounds. Using the residues as fertilizers is problematic because of such high concentrations.

Water is a major issue. When I was involved in HogWatch Saskatchewan, one of our members became a fervent activist because the major CAFO corporation had bought up rights to a significant guaranteed portion of her small community's municipal water supply for a full two decades. Beyond that, there are the very serious problems of water pollution that frequently occur with hog CAFOs.

Regarding communities, consider the fact that, compared to the past, the vast majority of profits leave the community and end up with anonymous shareholders. The trend toward CAFOs is among many that lead to massive rural depopulations and the loss of significant tax bases to supply proper services. The corporate pork industry has a lot of economic power. Many states have passed "right to-farm" laws which prevent counties from holding plebiscites to prevent CAFOs and have laws that prevent "disparagement" of their operations and make it illegal to film in hog barns without permission. American land grant universities are complicit and take large amounts of money from industry.

Conclusions

A UNESCO policy brief (Steinfield et al. 2008) effectively summarizes the global pressures for meat production. It is driven by market demands for much more animal protein. The drivers include rapid urbanization, rising incomes, longer global food chains, technical innovations, free trade deals, and highly variable grain prices that often favor their use for animal feed. The report concedes, in an attempt to be fair-minded, a few potential positive impacts including more better-nourished people. On a global basis, livestock production is a huge sector employing 1.3 billion people, providing livelihoods for one billion of the world's poorest people. It constitutes 40% of the global gross domestic product in agriculture.

There have been trends to shift much of the production from the global North to the South—from temperate to tropical and subtropical regions such as the Amazon and also the use of more intensive forms of raising livestock. There are still large areas such as in Africa where the poor make use of extensive methods of livestock raising such as with pastoral nomadism. The global production has tripled from 1980 to 2002 from 47 million tonnes a year to 139 million tonnes and that is expected to double again by 2050.

Enormous pressures are being placed on the environment. Almost 30% of the Earth's ice-free surface is devoted to livestock production compared to 8% for the growing of foods directly eaten by people. As more emphasis is placed on intensive factory feeding of animals, even more pressure will be placed on croplands for animal feed. Biodiversity is being rapidly lost in tropical regions, and grazing is seriously degrading 20% of the total land being used for livestock production. There are extreme issues of both depletion and pollution. Livestock raising has contributed to profound changes in global "biogeochemistry." Nitrogen is released in huge quantities as ammonia and methane—60% of global anthropogenic ammonia contributing to acid rain and acidification of ecosystems. Livestock production produces 18% of greenhouse gas emissions—9% of anthropogenic carbon dioxide release and with 37% and 65% of anthropogenic nitrous oxide having an even greater potential to warm the atmosphere.

Some other impacts include contributions to over-fishing to provide high-protein animal feed. There is the elimination of biodiversity as the lands devoted to pasturing and animal feeds increase, along with the further loss of natural grasslands that have many important functions such as much more biomass, deep root systems with much better water conservation capacities, and a serving as important sources of carbon capture. Desertification follows in many cases. Too much biotic homogenization follows as with the vast majority of modern agricultural systems.

Human health factors are of major concern. Animals in concentrated feeding operations can be hosts or vectors for new and virulent strains of diseases as has been the case throughout humanity's domestication of animals when you consider the major infectious killers of the past—smallpox, tuberculosis, mumps, and measles. New strains are emerging with poultry and swine flu as examples exacerbated in their sudden spread because of high-speed rapid transportation that globally connects us all. Heavy diets of beef are associated with chronic conditions of cancer, heart disease, and diabetes. The heavy preventative use of antibiotics in feed creates a problem for humans in dealing with new super-strains of bacteria that evolve in antibiotic resistance. It adds enormous costs and vigilance to food inspection for safety and an ever-alert international disease control system. These systems are often put at risk because of the neoliberal and corporate tendency to pressure for deregulation and government austerity that reduces their effectiveness.

Social impacts abound. Many indigenous peoples and small landholders in tropical and subtropical regions have been brutally displaced to extend cattle grazing grounds or vast soy plantations to feed animals. The ever-growing vertically integrated corporate system displaces many thousands of formerly independent livestock producers in Canada and the United States contributing heavily to rural depopulation and poverty, and communities can become bitterly divided over the siting of massive animal operations in their regions. Work on these operations tends to be low paid, highly stressful, generally unhealthy, and with high turnover. There is often a heavy reliance on undocumented, temporary guest migrant workers who may be at the mercy of unscrupulous bosses and without the social protections given to local people. Conservative permanent residents may resent their presence, leading to other types of social conflicts.

Certainly, significant shifts in diet toward more plant-based protein, and more sustainable ways of raising livestock would go a long way in confronting the environmental issues we face. Meat production clearly is a major political ecological issue.

5 Who Really "Feeds a Hungry World"?

It has been the contention of many that the only way to feed a hungry world is through the Industrial Food Chain. That claim is amplified by those suggesting that only a Second Green Revolution involving genetically modified crops will be able to feed several billion additional humans. Are they right?

Development economist Raj Patel (2009b) strongly disagrees. These systems have had over one hundred years but have exacerbated social disruptions and inequalities while making their own environmental messes and health crises such as pollution, greenhouse gas emissions, obesity, diabetes, cancers, and pesticide poisoning. Additionally, the several billion who face poverty and hunger and malnutrition, industrial farming have not seen their problems resolved because they suffer from unequal allocations of food.

Considering protein, carbohydrates, and other nutrients, far more than merely enough is now produced, but those who generate the long value chains are interested in where the profits are made—not really in "feeding a hungry world." Major agricultural powers such as the United States tend to use food aid as a geopolitical weapon to influence others to acquiesce to its ambitions rather than solving long-term food insecurity. So much allocation of grains, most especially corn for livestock feed and processed products such as soft drinks and junk food, shows lack of commitment to global food security. The neoliberal attempts to extend free trade have neither contributed to food security nor to well-being. Patel (2009b) uses the Mexican example where cheap corn flooded in from the United States has led to major displacements, increased poverty, and poor nutrition. Farmers in India have been made extremely vulnerable to food price fluctuations through trade liberalization while facing huge drops in rural income and epidemics of farmer suicides. Soy plantations in Brazil and Paraguay, intended as a cheap source of animal feed, ultimately benefit affluent consumers living elsewhere and internally lead to the displacement of many thousands of landless agricultural workers and the exacerbation of poverty.

Government policies throughout history have controlled food. The Romans provided grain to keep the mobs quiescent and Britain changed its corn laws in the 19th century to provide cheap grains for industrial workers while keeping wages low. It drew upon the hinterlands of the Empire for inexpensive wheat, beef, salmon, and stimulants such as tea and sugar, provisioning its homeland workers.

The United States used a mercantilist approach of protection, subsidies, support systems, and tariffs to build up an industrial farming system that simultaneously supported the building of its massive manufacturing system. Then after World War II, it used its agricultural might to dictate a global economic order designed for its own benefit. While such national food policies were always ultimately to the favor of ruling elites, they did to some extent benefit classes lower to them.

Alarmingly the role of fundamentally setting food policies has now shifted to the private sector in the form of transnational corporations. International financial institutions such as the World Bank and International Monetary Fund, at the service of the neoliberal order, have been dictating food policies to developing countries that became deeply in debt by those very same institutions. Countries have been forced to relinquish attempts at food sovereignty and, in the name of "comparative advantage" and "free trade," produce commodities for sale in the developed countries and permit foreign corporate land grabs. At the time of Patel's writing over 40% of world trade in food was in the hands of a few transnational marketing companies such as Cargill and Archer Daniel Midlands and is presumably now much larger. They do this in alliance with biotech seed and pesticide companies such as Bayer/Monsanto and Syngenta. Together they powerfully lobby their national governments and ultimately generate the trade policies regarding globalization and food that their home countries promote on their behalf.

There have been counter-movements through the emergence of small-scale farmers and peasant organizations. They have arisen in almost every country, including the United States, Canada, and in Europe but most especially those under the most severe threat in the Global South. Central to these movements has been the concept of *food sovereignty*—the right of a people to collectively achieve a nutritious diet, in environmentally sound ways, based on local conditions, and to determine all aspects of their national and regional food policies.

Tied to that has been the emergence of a scientific field known as *agroecology*. It focuses on non-chemical ways of restoring and preserving soils; bio-natural ways of pest control, and encourages diversities in cropping that match the best conditions of local environments. It promotes local social reforms such as enhancing the role of women, and privileges cultural factors—values, preferred foods, and ways of cooking—coordinating them with any proposed innovations. Agroecology is participatory and relies on partnerships with peasant knowledge and practice.

The term peasant may imply "traditional," "conservative," and "resistant to innovation." It does not mean that here. Peasant simply means that the producer devotes a significant portion of their crops and animals to feeding their family and is tied closely to their own community while seeking some income from market sales. Historically, the vast majority of farm innovations began with peasants. It is only very recently that we have turned the responsibilities over to scientific specialists who themselves do not farm.

Industrial Food Chains vs Peasant and Small-Farmer Food Webs

The research organization ETC Group (2017) outlines a set of contrasts in keeping with the contentions of this chapter. *Peasants embedded in their local webs are still the world's major producers of food.* They feed more than 70% of the global population, yet they only do that with about 25% of the resources of land, water, and fuel. In contrast, the Industrial Food Chain, with 75% of the world's agricultural resources, provides food for less than 30% of the globe's population *yet is among the leading sources of greenhouse gases.*

The cost of external damages created by the Industrial Food Chain, ETC claims, is quite expensive—twice that which society pays to retailers for the food. *The claim is also that the collective behaviors of the industrial food system, along even with its research and development capacities, are seriously deficient in their capacity to respond to climate change that it in fact has quite greatly exacerbated it in the first place.* In contrast, the more localized peasant food webs tend to support biodiversity. This would include plants, livestock, fish, and forest in broad spectrum, and many particular species, most notably pollinating insects not harmed by insecticides.

What is the Peasant Web? It consists of small-scale producers, often female-led, and can include urban and peri-urban producers as well as farmers, livestock producers, hunters, and fishers. They may combine seasonal farming and work at part-time wage labor. They may sometimes produce for the Industrial Chain and occasionally consume from it. They may or may not be self-sufficient in their production but they usually trade or sell some of their produce to local and national markets and grow some crops—coffee or tea, for example—for international chains. The web is not necessarily synonymous with organic agriculture since a few may use some pesticides and synthetic fertilizers in their operations, but generally their production is largely organic due to farmers using tried and traditional ways of doing agriculture and not bearing the costs of machinery and chemicals.

The Industrial Food Chain is linked in all its dimensions. The first link is within the research and development sector where plant and livestock genetics are of primary concern and where new varieties of crops and animals are not bred at farms but at government or private labs and experimental farms. Similarly, pesticides, machinery, fertilizers, and veterinary inputs are devised at off-farm locations and their uses are influenced by government policies and corporate ambitions for profit. Within the Chain and while still always vital the farmer plays a comparatively minor role in decision-making. Essential commodities are still produced there although that is not where the ultimate profits end up or the important decisions are made. Beyond the farm there are many links of transportation, storing, milling, value adding, and processing before the commodities are ready for consumption. When ready, they are delivered to wholesalers and then to retailers, restaurants, and homes for consumption. All aspects of the Chain are tied to the market economy and to political and financial systems—future markets gambling on profits by nonfarmers, and government export policies very much influence any actions in the chain. All this leaves the farmer and his or her community exceptionally vulnerable and with ever-lessening control over their destinies.

The estimate in the report is that in 2017, 4.5 to 5.5 billion of the world's population get their food from peasant webs. Three and a half billion rural people are direct participants in it as producers and consumers primarily in the global South, but there are still millions in the North that participate as peasants or in community-shared agricultural cooperatives. ETC calculates that about a billion urban food producers make heavy use of gardens, small-scale animal husbandry, and fishponds. Hundreds of millions are drawn into the peasant webs when the commodities in the Chain are unavailable or too expensive. Seventy-seven percent of food crops and livestock production is still consumed within the countries where they are produced and most of that is gained from the Web. The exception would be highly developed countries where the Industrial Chain dominates and where much of the food is traded on a global market. Peasant farmers in the Global South still harvest the majority of calories—for example, 80% of rice. Urban agriculture, although largely absent from developed countries, is fast becoming highly relevant in the urban South with street vendors selling garden produce, and chicken and eggs feeding 2.5 billion people to a large extent.

ETC characterizes the Industrial Food Chain as highly wasteful—76% of its total calories are directed elsewhere before making it to the human plate. Humans directly eat only 24% of it. A huge portion of the Chain's calories is devoted to meat production. Considering calories from field crops, 9% goes into biofuels and other nonfood products. There is loss in the complicated transportation, storing, and processing aspects of the chain—ETC's estimate being 15%. Households in the more affluent countries waste another 8%. In those countries, there is a tendency to over-consume, as is the case with products processed from corn. *The Chain does not serve the globe's rural poor because they are too remote and too poor to provide profit for the market dominated logic of the chain.* There is a massive imbalance in resources in producing this food for which there is so much waste. The chain uses 90% of agriculture's fossil fuels, 80% of its water use, while monopolizing 75% of the world's agricultural land. In that process, they estimate that 75,000,000,000 metric tonnes of topsoil are destroyed every year and much deforestation continues.

Research and development dollars have overwhelmingly gone to crops in this sector. Maize alone receives over 45% of the private research and development funding and that which is becoming dominant in the chain—genetically modified crops—is enormously expensive. For each GMO variety alone, it takes an average of $136,000,000 to get to market.

The research and development that is devoted to the Peasant Web is negligible, although the report sees a great deal of promise through the field of agroecology (see Chapter 11). Here scientific agronomy works in equal standing with peasants and their knowledge in long-term, ecological, polycrop responses to gain the best overall productivity while preserving soils and reinforcing resilience. The ETC report points out that peasants have domesticated over 7,000 plant species and have generated from that over 2.1 million varieties! From that massive gene pool humanity has been and will always be dependent, peasants tend to save seeds and share and locally trade them without buying them from the Chain.

A serious issue arising from the Industrial Food Chain is that of human health. Sixty percent of all human diseases come through domesticated animals that are kept in huge and concentrated numbers in CAFOs. Diseases such as avian flus are reinforced by the genetic uniformity among the birds. In contrast, those in the peasant web maintain considerable genetic variation. About $24 billion is spent annually on pharmaceutical sales by livestock corporations most significantly as antibiotics, which are also used as growth promoters. Antibiotic resistance has become a major health issue and ultimately threatens pandemics of yet unimagined proportions.

Corporate power over farmers and ultimately peasants is a major problem. Three companies dominate the global seed market worth $55 billion annually, now pushing ever-increasing GMO varieties after they bought up over 200 smaller companies. Pesticide companies linked often to the raising of GMOs are again dominated by three that control over 51% of a $63 billion industry. Their concentrations of power are beyond that which even most nation-states can control. Fertilizer sales in the form of nitrogen and other synthetic forms are similarly massive, reaching over $175 billion annually. Only about half of the fertilizers reach the plants, and there are huge amounts of environmental damages to soils and waterways.

Insect pollinators are yet another victim of the Chain. Wild insect pollinators numbering more than 20,000 species supplemented by birds and bats tend to be protected through traditional approaches. Yet through the Chain, much "collateral damage" is done to highly useful insects that provide a collective service that is valued at over $235 billion a year. Beyond that, pesticides along with synthetic fertilizers are killing off valuable microorganisms that maintain soil functioning.

Obviously fossil fuel use is exponentially greater in the industrial chain than in the peasant webs. The ETC report suggests that it takes nine times more energy to create the same kilogram of rice industrially as compared with the Web. Half of the energy used in growing wheat by the Chain is used to manufacture crop fertilizers such as nitrogen and pesticides. They assert that "the average American uses 2000 liters of oil equivalents per annum to put food on the table" (ETC Group 2017: 35). Preservation and packaging are a major part of the Chain's strategy for long-lasting and attractive marketing. Agriculture through the Chain is responsible for significant proportions of contemporary greenhouse gas emissions and one-third of that is due to livestock production.

Overconsumption can be blamed on the Chain. Americans eat more than 25% of what they need due to an agricultural system that, through the use of subsidies, has fostered the production of high-calorie, largely processed, cheap food. Globally, the number of obese now numbers 30% of the population—for the first time surpassing those that are hungry. The direct food bill paid by consumers is $7.5 trillion—but that includes $2.49 trillion lost or wasted along the Chain. They consider $1.26 trillion as the price of overconsumption. Then their calculation for the social, environmental, and health costs is $4.8 trillion. So, the total cost in their calculation is $12.37 trillion, suggesting $8.6 trillion beyond the cost of the food consumed, meaning, in their view, 69% is counterproductive.

The Chain has a dismal record when it comes to social and cultural impact. Cultural diversity is parallel to biophysical diversity—the more variation, the more the sources of adaptability and resilience. As the Chain, a standardizing and colonizing force, moves its way deeper and deeper into the global South, more languages and cultures are lost that carry vital information capable of accessing the indigenous knowledge of the land. Monocultural food systems disconnect consumers from farmers while changing our dietary customs and food choices to adapt to the Chain that provides fast and processed food in ever-increasing quantities that are separated permanently from tradition. In spite of this being an era labeled as the "Information Age," as they put it, "our generation may be the first in history to lose more life-supporting knowledge than it gains" (ETC Group 2107: 43). *The Web by its very nature respects and nurtures diversity in all aspects, including the socio-cultural.*

A period of two centuries of transformations caused by the Chain has led to the elimination of hundreds of millions of family farms, leading to an agricultural sector that in the developed countries only employs 50 million people on modern farms and has driven most rural peoples to the cities. This Chain system greatly threatens livelihoods in the South where people are being divested of their lands to make way for extremely large farms and plantations or development projects such as dams and reservoirs that support agriculture. Insofar as they might eventually be absorbed by urban manufacturing, there have to be some limitations. Even if it were possible to expand those sectors indefinitely, it would only lead to more and more cheap exploited labor subject to lax regulations.

La Vía Campesina and Food Sovereignty

This movement described by Desmarais (2007) and Wittman et al. (2010) provides a direct counterpoint to industrial, corporate-dominated agriculture. La Vía Campesina (LVC), translated as "the way of the peasant or the small landholding farmer" was formed in 1993 at a time when international free trade agreements were rapidly being shaped and the World Trade Organization was about to be formed. This threat along with the alarmingly fast invasions of corporate agribusiness into the developing worlds was upsetting small land-holding farming that had fed the world for thousands of years and had accounted for the overwhelming majority of agricultural innovations that had served agriculture and all of humanity. This federation of global small farming organizations includes 149 organizations in 56 countries and represents approximately 200,000,000 farmers.

They created a network of solidarity and through direct farmer-to-farmer networks are building personal links and sharing innovations. While international in scope and activities, it emphasizes *sovereignty*—the rights of farmers and their communities to protect their local livelihoods and destinies. It resists top-down hegemonies of uniform development favoring corporate-dominated, costly input agriculture, and free-market globalization. It sees fellow farmers as having common class interests yet ones that are solidly embedded in local realities.

It has become a major and influential actor in agriculture and trade issues. In 1999, it became highly visible through its activities at the first meeting of the World Trade Organization and through the alliances it made internationally with trade unions, environmental organizations, and social justice movements. Then in 2003, due to its public demonstrations and effective lobbying, it was able block a proposal to globally eliminate *all* tariffs on agricultural products. Such a proposal, as demonstrated through the North American Free Trade Agreement (NAFTA) of 1994, showed what happens to small farmers who were displaced in Mexico by the massive flooding of cheap, highly American subsidized corn. If complete free trade covering all agricultural products were accomplished, it would impede developing countries' ability to feed their own people and destroy the livelihoods of many millions of farmers who form the backbone of both society and the economy. If totally non-tariffed trade were achieved, unfair disadvantages would eliminate local farmers and their products.

For instance, bananas might be imported more cheaply into African countries from Central American ones and the market might then eliminate local banana farmers. Yet transnational corporations, having the advantages of economies of scale, would have grown such bananas on large plantations backed by the cheapest labor from already displaced peasant farmers. In such a situation, social injustice would prevail through the elimination of African farmers, the labor exploitation of fewer farmers in Latin America, and the transfer of larger profits to the shareholders of major fruit companies most often headquartered in the United States.

LVC disputes that there is inevitability to the hegemony of the capitalist, industrialist, monocrop, globalized system. In contrast, it claims that peasants or small farmers still represent one third of the world's population and are still responsible for two-thirds of the world's food. Crises in global food circumstances are not due to shortcomings created by peasants and small farmers. LVC places the blame for the negative impacts on the global production chains dominated by the agro-corporations and the marketing chains dominated by supermarkets. Their incessant drives for profits benefiting distant shareholders, and the overemphasis on the production of animal feeds are more to the point in assigning fault. Corporate-dominated agriculture has generated greater levels of instability and hunger. This is due to the coupling of the agricultural system to the financial system that by nature generates instability in investment flows and production not tied to any morality of placing it where rationally and humanely it is most needed. During the 2008 Global Financial Crisis, there was a large shift of investment to grain futures that artificially and suddenly raised the costs of purchasing wheat, rice, and corn leading to much hardship and food riots. It was also exacerbated by a trend to invest heavily in biofuels that had taken up about 5% of the globe's valuable food-growing land.

Huge amounts of environmental degradation have been occurring because of industrial farming. Again, these include water losses and pollution, fossil fuel and pesticide contamination, carbon dioxide release, soil depletion, sterilization, and salinization, massive deforestations and losses of biodiversity, poor nutrition and obesity, and chronic conditions such as diabetes and cancers, and threats to human health with the emergence of new infectious diseases associated with CAFOs.

Consider the nature of foods that North Americans eat. The average item travels 1300 miles before becoming part of a meal. Fruits and vegetables are refrigerated, waxed, fumigated, irradiated, colored, packed, and shipped. These processes are done to enable distribution over long distances and to enhance shelf life—not increase nutritional value. Then there is the whole question of food safety arising from complicated chains of production and marketing. These concerns have been illustrated with the outbreak of "mad cow disease" that affected livestock production in Europe leading to the slaughter of millions of cows, the outbreak of Asian influenzas from the concentrated raising of poultry and hogs, and the outbreak of *E. coli* poisoning as the result of poor handling and inspections.

Since the 1990s, agribusiness with the aid of national governments, international aid, and financial institutions such as the World Bank, and university colleges of agriculture has been promoting the need for global *food security*. It is premised on the notion that the world's population is growing dramatically and that there are and will be "many hungry mouths to feed." This leads to a call for a second Green Revolution. There is little room left to extend the existing land base to feed another three billion people with current technologies and then given the severe consequences of environmental destruction, there supposedly will have to be radically new ways developed to produce huge amounts of food with new procedures. These arguments emphasize the coordinated research capabilities of national governments through their agricultural research facilities, universities, and, because of the huge capital investments and time required, the corporate.

The notion of food security also implies the maximum of trade liberalization through the free flows of food commodities and agribusiness investment. It doesn't matter where the food is produced—all that supposedly matters is that it gets to where it is needed at the best prices possible. Food security sounds benign and even humanitarian but is an extension of the neoliberal view of agriculture and all that is implied by that. It means extending the scientific and technological dimensions of green revolutions. It implies the maximum use of genetically modified/pesticide-ready crops and is based on the neo-liberal notions of comparative advantage and profit—in that each part of the world will maximize the crops that it grows best and export it to where it supposedly grows to a lesser advantage.

The alternative as developed by LVC is *food sovereignty, a much more comprehensive holistic notion and one that incorporates food security as a matter of social justice and within a moral economy.* Everybody should be nutritiously fed and not just by his or her ability to pay in the market. Justice and fairness dominate the notion of food sovereignty that covers peasants or small farmers, farm laborers, and consumers of agricultural products.

Food should not be treated as just another commodity. Instead, food is considered as a human right with deep cultural value, not something that should be seen just in the context of market logic. Food security logics, in tune with neoliberalism, emphasize individual and household choice. Altogether food sovereignty emphasizes solidarity and shared assets, mutual rights, and collectivity in ownership over resources. It places great value on things that are not easily quantifiable and outside of the market. Values are placed on culture, biodiversity, and traditions in knowledge and practice. They put a decided emphasis on localization—any control over production and distribution should be rescaled locally.

Self-determination within the state and between states is underscored in food sovereignty. Corporations have been in the process of taking control away from the nation-state. The state needs to reestablish the sovereignty of peoples and their communities in guaranteeing affordable prices for the consumer, but prices that justly compensate the small local producer. The state needs to be more aggressive in bringing about true agrarian reforms that deliver justice to small farmers and displaced landless rural workers. Trade in the international arena should always be subordinated to social goals.

Healthy good quality food that is culturally appropriate should be promoted—food destined for a domestic market in a quest for establishing as much local self-sufficiency as possible. That production needs to be based on diversified farm production rather than on monocrop overproduction. Food sovereignty is focused on ecological relations with local environments that work with nature in maintaining biodiversity and the long-term preservation of soils. It also requires curtailing of the ever-mechanization of farm production that generates impossible costs for most, leading to major debt and displacement from farming, and major contributions to CO_2 releases. The survival of the family farm is viewed as paramount.

Its logic asserts that the state needs to protect farmers, both men and women, with remunerative, stable prices, and protection against wildly fluctuating and low ones. This also means controlling production internally (supply-side management) to prevent excessive surpluses that are typically propped up by unnatural subsidies that allow such countries (e.g., the United States or the European Union) to flood international markets with their products putting small farmers out of business in nation-states that do not have similar political advantages. Food sovereignty requires the elimination of all direct and indirect export subsidies.

LVC is also adamantly opposed to the control of seeds and their distribution by transnational corporations as well as to GMOs and the patenting of seeds. Peasants have the proven means of conserving biodiversity through their own resources and traditional knowledge and methods of sharing. They have the right to choose, store, and freely exchange seeds and the genetic resources contained within them.

Conclusions

The world faces two major challenges regarding food and agriculture. One of them is the possible addition of three billion to the already dangerously large population of eight billion. We have run out of new farmland, but the malpractice, for instance, of deforesting large quantities of tropical forest to compensate as with soy plantations and ranching in the Amazon, has had devastating consequences. It has only exacerbated atmospheric imbalances and gravely damaged biodiversity.

The other challenge is climate change itself, and there is uncertainty about what changes will happen in particular zones and at what specific rates. We do not know which areas of the world will be suitable for which crops and at what level of productivity and which current ones will not and become degraded in their agricultural potential. Responding will demand a huge amount of human ingenuity for flexible, highly varied, adaptive, and resilient forms of agriculture.

Those positioned within and committed to the corporate, agro-industrial realm advocate staying the course with more of the same but in greater intensity and emphasizing engineered solutions coming from laboratory and experimental field solutions—top-down technocratic approaches. A major part of this is a confidence in a kind of Green Revolution Two—with an emphasis on increasing the range of genetically modified or transgenic crops that would supposedly increase yields and contain special traits adaptive to the new conditions generated by climate change. For the time being though, the only significant advantages of GMO crops over ordinary high-yielding hybrids seem to be in their capacity to be resistant to particular chemical pesticides or to specific insect pests such as corn borers. Some farmers, when they can afford it, are possibly able to manage their defenses against some pests with greater ease and efficiency.

There are many problems with such a reliance on transgenic crops—only to mention a few here. They require huge capital investments—sometimes a hundred million dollars or more for each variety (ETC Group 2017). Such investments can only be handled by very large corporations such as Bayer/Monsanto with their already tarnished reputations for environmental, health, and transactional ethics (Robin 2010). Profit and neoliberal motives rule supreme, and investors are anxious to rush their products to market. Then they gain earnings through sales to farmers by locking them into a kind of bondage of dependency on their bundled products (seeds and pesticides) as with Round-up ready corn, cotton, and canola. They expect as little testing and regulation that they can get away with from compliant governments that tend to be already strongly influenced by this corporate sector (Vallianatos 2014). Placing this much power in the hands of a neoliberal corporate sector is not in the best interests of farmers, consumers—and not our descendants.

Adherence to such a technology is quite expensive for the farmer, adding greatly to input costs of chemicals, machinery, fuel, and frequent irrigation. This further complicates the trend where farms have to get bigger in order to survive, thus displacing large numbers from the rural areas or forcing them to become poorly paid laborers for gigantic corporate farms that are not prone to consider environment over profit. We have documented some of the negative consequences for developing countries with the examples of ranching and soy plantations in Latin America. Would we want to extend such systems to the rest of the developing world?

Back to the transgenic crops themselves—the manufacturers draw, frequently in a piratical fashion by claiming intellectual property rights, from the vast genetic heritage of landraces entirely discovered by peasant farmers (Shiva 1997). However, by nature of their investment costs and searches for specific traits such as particular pesticide resistance, these GMOs promote mass uniformity, greatly reducing diversity, and leading to situations where farmers in very different zones use exactly the same variety of GMO. This has been a disturbing trend of high modernity and globalization—massive sameness that then reinforces the meta-assumption "that there is no alternative." What is required instead, some would suggest, is the maximization of diversity to meet the highly varied environmental conditions that will be rapidly changing. Could then the standard method of transgenic crop production result in a super-superior high-yielding variety that was, at the same time, drought resistant, tolerant of very wet conditions, extreme heat, and frost while being resistant to a particular set of pesticides that do not generate super weeds or super-bugs that can prey upon them? A tall order, impossible to meet, and parallel to those in the energy sector who suggest the building of thousands of expensive nuclear reactors is the only way to solve the climate and energy crises.

Beyond these issues that have been focused on field crops are those related to the production of animal protein—as associated with CAFOs and with the overemphasis on growing animal feed in spaces that are needed for growing crops to directly feed humans, plus the greenhouse gas emissions resulting from all of this. These issues are also in desperate need of solving.

Then there are the huge energy costs of the current agricultural/food system. González de Molina and his colleagues (González de Molina et al. 2020) use the notion of systems "far from equilibrium" where they place their perspectives within the interdisciplinary field of emergence and complexity studies. For any life or cultural system, energy must be captured and expended to maintain for as long as it lasts, and systems are open to some or a great extent to a range of others where energy, influence, and information flow among them. Yet, physical entropy is always generated in any process involving energy—as a breakdown into heat and smaller masses of particles that are dissipated as toxic waste or greenhouse gases. Clearly, the corporate food regime is open and subsidiary to the industrial capitalist system as a whole, thus making agriculture maximally dependent on the fossil fuel industry where entropy is colossal. So high entropy is the first characteristic through processing as gas and oil, fertilizers, and pesticides. Next, there are huge transportation, refrigeration, processing, and packaging costs through to the retail context. The consumer then buys the products at a big-box supermarket at the edge of her city adding more than the transportation expenditures through refrigeration and cooking. Obviously, this is altogether a huge energy expenditure accompanied by similar excessive amounts of waste, pollution, and high entropy.

So, there are major problems to be solved. As in the past when modernity created or reinforced social problems, social movements arise to counter them. This has been happening with agriculture. As with all movements, the main task has been to frame grievances and then to come up with ways to solve them. We have seen the rise of the La Via Campesina as a global federation of peasant and small farmer national organizations. This movement has found significant scientific support through agroecology as a rigorous discipline that blends ecological knowledge with agronomy, works with nature instead of pushing it, and recognizes the need for holism by taking into account social and cultural traditions. It too supports food sovereignty and localization.

Given environmental and socioeconomic problems represented through industrial agriculture, it seems appropriate to start somewhere else other than systems that have been exacerbating the problems. So, paying attention to agroecology (see Chapter 11) in the context of the hundreds of millions of small farmers that remain, especially in the global South, seems the logical thing to do. Yet to change circumstances, there have to be major changes to regimes of political economy (see Chapter 10).

Finally, a startling and essential need to reform our current global agriculture system has been revealed in an article in *Science* (Clark et al. 2020) by a group of international scientists who first point out that agriculture currently accounts for 30% of greenhouse gas emissions. Then they document how even if all our other energy systems were reformed to no longer emit such gases, the current practices in agriculture would prevent temperature rises staying below 1.5° and 2.0° Centigrade. Clearly, agriculture badly needs radical transformations that are ecological rather than industrial.

Part II
Energy

6 Coal

Coal provided the energy to bring about many positive transformations in civilization and it powers many of the electric conveniences that few of us could live without. Yet it has been at the heart of so many environmental, health, and social problems. Supposedly on the way out because of its clearly documented links to global warming and air pollution, it still holds the world's number one position for global electric power production—36%—in contrast to oil (2.5%), gas (23%), nuclear (9.5%), hydro (15%), solar (3.6%), and wind (6.5%) in 2022 (https://www.worldenergydata.org/world-electricity-generation/). I present here an overview about the impact of coal and then illustrate just some of the negatives with a case study—the shockingly, callous practice of mountain-top removal in West Virginia.

Coal led developed nations into the stage of high modernity. Coal was "King" and for the longest time was the world's largest industry along with its function as the dominant source of power for most enterprises. Eventually, coal was supplanted by oil and gas as the more dominant sources of power for transportation beginning and after World War II. Even then, abundant coal still served to maintain industry even at an increasing level.

While no longer as vital for transportation or manufacturing, coal has remained the largest global producer of electricity and is still essential for the steel industry. In cases of steel and electricity and beyond, coal remains "King" in countries such as China, India, South Korea, South Africa, and Indonesia, and dependencies on it are still growing in those regions. So, coal, in spite of international efforts to eliminate it, continues to rise in its notorious position as the number one emitter of greenhouse gases.

Overview of the Significance of Coal and Its Environmental, Social, and Health Impacts

The use of coal as a power source ushered in the Industrial Revolution. That reality has to be considered much more than simply a mixed blessing. Since the late 1700s, its combustion has released many billion tonnes of carbon dioxide into the atmosphere. Its mining, processing, transportation, and burning have also resulted in many more environmental and social consequences beyond this alarming predicament.

Scheidler (2020) compares coal's abrupt appearance in the context of human history as metaphorically similar to a sudden discovery in astronomy of a supernova filling a large section of the night sky. Regarding power, its consequences dwarfed everything that preceded it. Coal was the basis of a new kind of energy that was generated through industrial might the largest and most powerful political and economic entity the world had known to that date—the British Empire. That was accompanied by the ascendancy of capitalism, its extractive powers, and the extension of its reach globally.

One of the significant contributions of coal was its transformation into coke made through extreme heating without air, thus replacing charcoal in the smelting of iron and steel. It came at a time when forests that had provided the raw material for charcoal were dangerously depleted, and coal-rich Britain greatly benefited in that transition. Along with placing steam engines on movable platforms as machines for rapidly hauling new and larger kinds of carriages and carts on rails, Britain generated a huge magnification of its transportation capacities and then directly or indirectly stimulated the rest of the world—30,000 kilometers in Britain and 775,000 worldwide—much in British-held India by 1900 (Smil 2017).

While a major early factor in capital investment and accumulation itself, railroads enormously expanded the reach and capacity for delivering consumer goods and transporting people and livestock and supplying raw materials to factories where manufacturing was accomplished through stationary coal-fueled steam engines. The demand for iron produced through coal-based coke grew exponentially for the building of the rails to move the trains. Similarly grew the demand for steel to build locomotives, passenger cars, and rolling stock to transport goods and raw materials (Smil 2017). As part of the ever-growing, global network of transportation was the emergence of iron-built cargo and passenger ships powered by steam. Tied into all of these feedback mechanisms were ever-growing demands for coal let alone its uses in the manufacturing and domestic sectors and it became the common household basis of heating and cooking. Then in the later part of the 19th century, coal became, along with hydropower, the power source that initiated the global spread of electricity essential to almost all modern activities. MacKay (2009) notes that since the late 1700s its extraction has increased 800-fold.

Coal, in its dangerous working conditions, oppressing workers, oddly and indirectly contributed to some noteworthy levels of progress and social justice. Coal mining required large numbers of wage laborers working at particular locations and thus provided an opportunity for organizing particular forms of social movement and work strategies that eventually benefitted the majority, non-capitalist component of society. This was through the rise of militant labor unions and their powerful economic leverage of mass work shutdowns—or strikes.

As Scheidler (2020) points out, 19th-century capitalism and what he calls its "Megamachine" of power and social dominance were still absolutely dependent upon coal to keep virtually all of its industries running. The mining of coal provided a vulnerable bottleneck that if interrupted could stymie all industrial operations. The bloody history of labor strikes and unions in Britain and the United States was accompanied by much suffering. Yet in the long run, ushered in many of the societal rights and concessions for the well-being of the public through laws determining better working conditions, occupational health and safety, fair wages, and other progressive remedies that are in place today.

Interesting notes about the social history of coal aside, coal remains the main culprit in any concern about climate change. The International Energy Agency in March of 2022 reported that global carbon dioxide emissions had reached their highest levels ever in 2021—36.3 billion metric tonnes. Within that amount, coal represented 15.3 billion tonnes compared to 10.7 billion tonnes for oil, and 7.5 billion for natural gas. The main reason for the CO_2 increases was found in that year's significant upturn in coal usage (https://www.iea.org/news/global-co2-emissions-rebounded-to-their-highest-level-in-history-in-2021).

A team of researchers from the Harvard College of Medicine investigated the extraneous costs of coal and estimated if they were taken into account, it would double or triple the annual costs of coal annually produced in the United States. They calculated the price of the externalities was up to $500 billion a year (Epstein et al. 2011). It was pointed out that coal alone accounted for 41% of all CO_2 emissions globally in 2005 and, as we have just seen above, continues to hold that notorious leading position. Methane releases from coal mines also have a negative impact through their even more magnified greenhouse effects.

The costs, historical and current, were staggering. Coal mining and its combustion release into the atmosphere and physical environment many more and large quantities of toxic materials and greenhouse gases than just CO_2. They include lead, arsenic, cadmium, manganese, chromium, mercury, beryllium, nitrous oxide, sulfur dioxide, particulates of black carbon, and many micro air particulates of the PM10 and PM 2.5 (extremely small as measured in microns) and as mixed varieties that constitute serious health depleting issues of air pollution. Coal is the main culprit when it comes to air pollution and can be quite lethal in its degradation of water sources.

Since 1900, underground mining and its accidents have directly led to over 100,000 deaths in the United States—let alone the millions of injuries in such a dangerous occupation. Black lung disease killed over 200,000 people there throughout the same period. Of any industrial activity, workers in the coking of coal in preparation for the steel industry have the highest rates of lung cancer. Communities associated with coal mining similarly have markedly negative health risks in comparison to non-coal mining communities as shown by studies in Appalachia. These include lung cancer, heart, respiratory, and kidney mortalities, elevated hypertension, and low birth weights among many other negative indicators. Communities that are near or directly dependent on coal burning for power generation are subject to extremely poor air quality resulting from sulfur dioxide, black carbon particles, nitrous oxide, and PM2.5 particulates, and suffer serious respiratory illnesses at rates three to five times greater than those distant from coal-burning plants. Associated with these factors are increased cases of lung cancer, heart disease, many asthma attacks, and many thousands of extra hospital visits. These factors altogether represent only a few highlighted dimensions of health risks that are accumulative and ongoing.

Environmental destruction and degradation are considerable, especially with mountain-top removal or surface-stripping coal mining. Water sources may become seriously polluted with the release of heavy metals such as mercury, arsenic, and beryllium. Acid rain destroying forests and poisoning lakes and rivers comes from the releases of sulfur dioxide and nitrous oxide burning of coal. Overburden from mountain-top removal has destroyed many pristine streams and rivers in Appalachia along with water pollution and much biodiversity loss. Coal waste storage systems if broken or breached are extremely toxic to humans, water, and biodiversity. One example is post-coal mining cleaning slurry pits that tend to be elevated impoundments as located in places such as Appalachian valleys or valleys in the coal-mining regions of Wales. They are extremely hazardous and have occasionally burst or flooded over their containments killing hundreds while thoroughly contaminating neighborhoods while all of them are continuing risks to water quality through seepages.

After combustion in power plants, two types of left-over products must be very carefully buried. Coal ash is the heavier of the residues and fly ash is the lighter dust-like, toxic waste. Both are extremely dangerous. Fly ash causes cancer, reproductive, kidney, and neurological diseases, and diabetes. Over 1500 poorly constructed impoundments for fly ash are found in the United States and the risks are high for leaching into water supplies of contaminants such as arsenic, antimony, and selenium (Epstein et al. 2011). Bell and York (2012) report that annually air pollution from coal-fired power plants in the United States accounted for about 23,600 premature deaths, 554,000 asthma attacks, 38,200 heart attacks, 21,850 hospital admissions, and 26,000 emergency room visits. Other than nuclear radiation, it is hard to think of anything much worse than coal waste and post-combustion effluents, and, at least for now, the former is comparatively well controlled while the latter constitutes an ongoing and formidable danger.

A few brief notes about coal in China are needed to round out this overview. China is now the world's largest coal producer—accounting for about half the global production—and is one of the world's biggest importers of coal. It provides 70% of that country's energy in the form of heat, power, and electricity generation. Its mining employs around 10,000,000 people. Previously, mining was dominated by state-run corporations, but since the late 1970s many smaller local private firms have emerged. Tim Wright (2013) reveals three highly worrisome dimensions of the Chinese coal industry beyond its enormous contribution to greenhouse gas emissions.

First, there are large and worrisome amounts of corruption that overlap as significant causes for poor worker safety and environmental impacts. Because of the proliferation of many small private companies and a bureaucracy requiring the owners to navigate as many as eight layers of local government approval, bribery has been frequent and that spills over with regard to inspections of worker safety and environmental impact. On the part of the bureaucrats, the money exchanged is not always just for personal gain, but to afford some compensation for major shortfalls of revenues that were supposed to be provided by the central government to complete duties in delivering required services.

The second negative dimension is the abysmally low level of worker safety. According to Wang et al. (2011) China's death rate in coal mines is thirty-seven times that of the United States! Much of the reason is because of frequent underground mine collapses and accidents involving machinery. Wright (2013) reports that the death rate averaged 6,000 a year since 1990. Historically, quantity in production has always been emphasized over worker safety. In the more recent open era, labor was, at first, cheap and easily accessible because of the large migration from rural areas after the close of agricultural communes. Also, the market competition on coal prices was fierce, so owners tended to ignore safety costs in order to finesse more profit. These circumstances have been recently improved though, because the labor availability has shrunk, giving workers more leverage in demands for safety and wages; mining companies have been on a trend toward consolidation and the state is able to manage and enforce more inspections and regulations. But there is still a long way to go to improve occupational safety.

Environmental degradation and pollution represent the third trend associated with China's excessive coal use. The mines produce five billion tonnes of waste each year—much of it toxic and encroaching on agricultural lands while polluting rivers and groundwater. Since the Chinese economy depends on coal combustion, cities are heavily polluted with extremely low air quality. Of the World Bank's list of most air-polluted cities, twenty of the thirty are in China. Concern has been raised about the high use of coal for heating and cooking in domestic settings and the poor health consequences. While there are policies in place for eventual uses of greener technologies to reduce the enormous CO_2 production and environmental and health implications, Wright (2013) believes that it will be difficult to fulfill because of the intricate and complicated vested interests in coal that have been the basis of China's rapid and ongoing economic ascendancy.

West Virginia and Mountain-Top Removal Coal Mining

This case study involving the most destructive form of coal mining—that of mountain-top removal—represents one of the very worst-case scenarios in political ecology. It brings together massive destruction of the landscape, pollution on a huge scale, much loss of biodiversity, and social injustice to the extreme, along with dire social and health consequences for the local inhabitants. It represents a callous use of political and economic use of power by outside corporations causing damage to local people and their environment. Beyond these factors are the obvious continuing contributions burning of coal as the biggest source of anthropomorphic climate change. Sociologist Julia Fox in her (1999) review of its practice and condition refers to her study region, southern West Virginia, as an environmental *sacrifice zone*. A sacrifice zone can be characterized this way, "In such a space, the physical and mental health and the quality of life of human beings are compromised in the name of 'economic development' or progress—but ultimately for the sake of capitalist interests" (Lopes de Souza 2021: 220).

Underground coal mining in West Virginia began in the late 1800s with coal companies and railroads moving into a wilderness occupied by resolutely independent Appalachian mountaineers. Living in mountain hollows and narrow valleys, they practiced small-scale, largely subsistence-styled agriculture including raising hogs and distilling whiskey along with hunting and gathering in the forests. Soon much of the land was then quickly bought up by absentee owners from the coal industry that was then essential to the then very rapid expansion of American economic might. Rural industrialization brought rapid change and the mountaineers were drawn into coal mining jobs as were Afro-Americans migrating from further south and recently arrived Eastern and Southern European immigrants (Nida 2013).

When mines were established in the remotest southern region, the companies in most cases had to build brand new settlements—company towns—and coal camps of a smaller size. This was done to attract miners and their families and provide them with housing and domestic and community facilities. The conditions were draconian though. Workers had to sign what were called "yellow dog contracts" vowing to never join or engage in union activities and only buy at the company store where costs were often purposely higher than at regular stores. Rather than cash, they were paid in company script, which could only be used in company stores. As well the miners and their families sometimes found the price increases there practically swallowed up any wage gains that might have occurred. Miners were paid by the ton—40 cents at the beginning—but tons were "long tons" of 2400 lbs compared to 2,000 lbs at union mines in Pennsylvania. Camps, company towns, and mine entrances were patrolled by intimidating armed guards on company payrolls and managed by private detective agencies using brutal tactics to keep people in line. This oppression set the stage for what may have been the most violent class war ever in the history of the United States—coal miners and unions versus the coal mine owners, in collusion with state and federal officials and the courts (Nida 2013; Duafala 2018).

The United Mine Workers Union had struggled throughout the early part of the 20th century and into the twenties to unionize the region. Any miners attempting to join were evicted with their families from company homes and forced to live in tent camps. The conflict became violent on both sides with beatings, murders, assassinations, property damage, and armed attacks on miner camps running rampant. Miners, union officials imported strike breakers, militias, state troopers, local sheriffs and deputized locals, along with detective agencies and hired thugs, West Virginia national guardsmen, and even U.S. Federal troops all became involved in these bitter coal mine wars. This all climaxed in 1920 and 1921 with the Matewan Massacre and the Blair Mountain Battle between 7,000 armed miners and 3,000 anti-union deputized resistors. In the latter case, the miners were marching to free striking miners that had been imprisoned in the southernmost county of West Virginia. After three days of fighting and after the arrival of federal soldiers, they surrendered because of their loyalty to their country and since the majority of the miners were themselves veterans of World War I (Duafala 2018; Nida 2013).

Both events—Matewan and Blair Mountain—dramatically underscored an extreme case of class conflict and capital versus labor. It was not until 1933 with the passage of the National Industrial Recovery Act that West Virginia coal miners and American laborers in general gained full rights to organize and collectively bargain for better salaries and working conditions. This also brought with it the freedom of speech that the pro-union miners in West Virginia thought was already guaranteed through their country's Constitutional First Amendment—but one that had been suppressed by the mining companies and the State of West Virginia.

West Virginia has traditionally always been always near or at the bottom among the poorest of the American states with low education levels, negative health indicators, continued violent labor relations, dominating company towns, and coal industry-dependent people living on the margins of poverty. This is mainly caused by reliance on a single industry, and the dirtiest and most polluting of them all. Coal mining is still controlled by corporations headquartered in out-of-state cities such as New York, St. Louis, Philadelphia, and Boston where interest in the well-being of West Virginians is minimal.

Mountaintop removal (MTR) is a form of strip or surface mining that has been in existence since the 1970s, but, ironically, as an environmental policy, the U.S. Federal Clean Air Act of 1990 gave it a burst of rapid expansion with the strict mandate for low-sulfur coal. The counties in southern West Virginia contain the lowest sulfur content coal in the United States. This led to the region becoming even more of a sacrifice zone than it had once been through underground mining.

Rather than tunneling for coal underground, MTR involves using dynamite blasts and huge earth and rock moving machines to extract about 500 feet of mountain depth to get directly at coal-rich seams. This removed material is labeled "overburden," which indicates the callous, utilitarian, and profit-oriented ideology involved. The blasts are made with mixtures of ammonium nitrate after extensive and complete clear-cutting of the dense and highly biodiverse forests found in the region, topsoils are removed, and rock structures demolished and removed by the gigantic truckloads. The scale is massive as much as 17% to 25% of the mountains removed in some regions.

The mining is highly capital intensive ultimately for extracting coal relying on single machines called draglines as high as the equivalent of 20 stories and that can cost as much as $100 million each. The dragline can remove 110 cubic yards in one scoop. Drag-lining coal produces 6 tonnes compared to 2.5 tonnes per labor hour with underground mining. After the coal is separated, gigantic trucks hauling as much as 380 tonnes each take the overburdens and dump them in nearby valleys with their waste levels as much as 1,000 feet wide and 500 feet deep. The process is known as "valley fill" with the end result being that the flattened topography of the regions resembles sterile moonscapes (Fox 1999).

Consequential to the flattening of landscapes, flooding has become much more frequent since the networks of streams and rivers to sustain the spring-run-off have been covered with the valley fill. In just three years 1995–1997, thirty floods were recorded in MTR regions. In 1972, a coal-cleaning slurry pond flooded from the excessive rains and killed over 125 people. Thousands of miles of streams and rivers have been permanently destroyed along with many life forms including, as one example, rare insect species and their larvae that are crucial to the local fish populations in one of the most biodiverse regions in the United States. The rivers are polluted with iron sediments, manganese, aluminum, and sulfates. Acids destroy aquatic life with 76% of West Virginia's rivers considered polluted. It was calculated in 1999 that 450,000 people were without safe drinking water. Hundreds of thousands of acres of rich hardwood forests have been destroyed and any restoration of lands cannot allow for their revival as forests since the soils stripped of original topsoils remain too compacted and covered with lime. They tend to "recover" as grasslands with scrubby bushes (Fox 1999).

The health of people, both physical and mental, in the region is in marked decline. Pollution from MTR has increased rates of hospitalization, pulmonary disease, hypertension, kidney and heart disease, and cancer. Polluted drinking water increases rates of cancer of the liver, spleen, and digestive tract. Birth defects and low birth rates are on the rise. Negative mental health indicators include higher rates of substance abuse, traumatic stress symptoms, anxiety, depression, and insomnia, and are higher than in non-coal mining areas (Fox 1999). Marberry and Werner (2020) report higher cases of opioid use in MTR counties in Appalachia, which is noted for the highest rates of abuse in all of the United States and is broadly associated with social despair, poverty, and precarity. In their discussion of health (Cordial et al. 2012) the authors (drawing from Albrecht 2006) refer to a poignantly sad psychological syndrome characteristic of the region—"solastalgia." This implies a deep sense of loss among people who remain in a devastated sacrifice zone as in this case, a kind of nostalgia accompanied by a sense of dislocation and severe loss about an environment that once was but no longer is to be. In effect, *it is the feeling of being homesick without ever leaving home.* Considering how beautiful this region of Appalachia had once been with its rolling hills and mountains, magnificent hardwood forests, and pristine mountain streams, it is easy to feel empathy with the grief felt by Appalachian residents faced with the excessive lunar-type landscapes created by MTR.

The southern West Virginian counties where MTR is practiced are dominated not only economically but socially by the coal companies as reported by another sociologist—Shirley Burns (2007). As mentioned above, they built company towns controlling almost all aspects of community life with, among other things, company stores and very strict leasing agreements for the housing that they provided miners. Again, miners and their families could get evicted if they became "out of line." While a miner could potentially change jobs and move to another mining location, the pattern of company towns controlling towns was repeated everywhere so there was no local escape. In spite of the eventual normalization of union membership, the pattern of control became more dominant throughout the twentieth century with the consolidation of mining companies into a few gigantic firms.

Because of MTR, mining companies no longer need large numbers of workers, but ironically instead require the hills and valleys where farmers, retirees, and former miners and their families live. There has been a large reduction in the number of miners because of the highly mechanized nature of MTR. People and their communities are considered as standing in the way of MTR practicing companies that now employ capital-intensive means to extract massive quantities of coal to satisfy the market.

Where MTR is practiced, the blasting has threatened people's housing with damage such as broken windows and cracked foundations from the detonations, as well as noise and rampant black dust during the mining. This disruption is continuous with coal and overburden carrying trucks daily passing by in large numbers on narrow inadequate dirt roads where accidents are frequent with local residents. Legal recourse is limited largely due to state laws favoring the industry, and because potential plaintiffs are reluctant to take on the companies in court. Coal dust saturations and concerns about impoundment slurry flooding are of chronic worry for local elementary schools where drills for evacuation in case of their occurrence are standard. A 1966 accident at a coal mining town in Wales with a sludge overflow killed 116 children at a school—so the worry is very real and as mentioned above episodes of slurry flooding with major death tolls have previously occurred in Appalachia.

Because of the large losses of employment and outmigration of working-age people, the local populations have high proportions of the elderly and disabled—consequently there is much rural poverty. Those that have decent pay are those that work in the fewer jobs remaining in the coal industry, or secondarily in the school system, or welfare systems supporting the poor, aged, and disabled. Because most of the few well-paying jobs are found in the coal industry, those workers having them strongly support the industry in spite of its quite obvious environmental, medical, and social impacts. The same is relevant for companies such as machinery repair that are auxiliary to the industry (Burns 2007).

All of this has led to what another sociologist Shannon Bell (2009) has referred to as an enormous drop in social capital. In turn, this gap helps to account for the difficulties in generating any sustained local resistance and a general passivity toward legal action in registering the many complaints that could be registered about property damage, poor health, and continued daily upsetting of life quality conditions. Bell (2009) through extensive interviews found that social capital through trust of neighbors, reciprocity, and cooperative patterns was sharply on the decline in southern West Virginia communities. Communities overall are slowly dying with much outmigration of the young and better. The southern counties lost almost half of their coal mining workforce even though that is where the most intensive and productive mining occurs—from 32,139 employees in 1970 to 17,045 in 2003. Housing is on the decline because of MTR activities—Burns (2007) gives an example of the assessment of one home falling from $144,000 to $12,000. Some people doggedly hang on because of attachment to home place and nostalgia for where they and their ancestors have lived for several hundred years. But the pressure of the coal companies' actions persists in killing the communities off frequently by buying up neighbors' property at bottom prices due to much lowered assessments created by the damage of MTR in the first place—a continuing vicious cycle of decline, degradation, and displacement. Ridding the region of people seems to be at the very least an implicit policy of the coal companies (Burns 2007).

Burns summarizes the appalling circumstances of southern West Virginia,

> In less than two decades, MTR caused irreparable harm to the environment, the culture, and the people of West Virginia. If coal companies continue to operate outside of the laws, if politicians continue to allow coal interests to wield ultimate power, and if MTR is allowed to continue unabated, then the coalfield communities will dry up completely. Ghost towns will spring up throughout the region. Grasslands will replace the hardwood forests. Moonscapes will replace the gently rolling mountains. The wilderness will be gone. If MTR is not legally halted or diminished, southern West Virginia coalfield communities, the people and land will be gone as the last ton of coal scraped out of the mountains themselves.
>
> (Burns 2007: 143)

An appalling footnote to all of these circumstances and the history of labor's struggle occurred in the 2010s when two mining companies initiated a court action to get the official status of the Blair Mountain Battle location as a National Historic Site removed. The mine companies wanted to establish MTR on that site (Mother Jones, Nov. 10, 2010, https://www.motherjones.com/environment/2010/11/massey-arch-coal-blair-mountain/). The memorial site is of enormous importance and pride to miners, their descendants, and families because of the heroism and determinism symbolized in the struggle for labor rights The companies were successful at first but fortunately a social movement and a later court appeal reversed that judgment permanently in 2018.

Conclusions

Considering the prosperity that coal mining brought to its owners and other sectors of industrialized society, the miners who do the highly dangerous work of extracting it have seen little benefit. They along with their families and communities paid a large price with their health and living conditions. As Jencks (1967) reports, British coal miners early on suffered from disdain and low social standing in the country's rigid class system. In Scotland's 16th-century coal mining, they were actually at first slaves and then serfs well into the 18th century in all of Britain. Even with general improvements in social conditions, because of their social isolation and low status, miners' attempts to gain improvements in wages, along with working and living conditions, greatly lagged the rest of society. Attempts to improve them were dashed during long periods of depression, poverty, and unemployment. It was not until 1946 when the Labour Government nationalized the industry that the miners could report some significant improvements in both social standing and working and living conditions.

Considering the history and circumstances of coal mining, it would be hard to avoid the conclusion that coal miners, their families, and communities may have suffered among the worst social injustices pertaining to the working classes. Morrice and Colagiuri (2013: 75) define social injustice "as the unequal or unfair social distribution of rewards, burdens, and opportunities for optimizing life chances and outcomes." Coal community research reports poor quality water for drinking and use in gardens, deteriorating conditions for outdoor recreation, dingy and uncomfortable clothing even after washing, and the frequency of having to use filter systems or bottled water. Extremely deteriorating health outcomes markedly affect miners and members of their communities compared to non-coal mining communities as has been shown.

Even today, there has been a major resurgence of black lung disease in Appalachia, specifically with MTR in West Virginia. This chronic affliction is associated with the silica particles that have more recently been discovered, along with coal dust itself, to be an important part of the disease's etiology. The silica portion is now much higher with MTR because of the massive crushing of rock overburden where it is always present in large quantities. Bodenham and Shriver (2020) expose how mining companies avoid safety measures, deny and minimize the dust's presence, falsify records, contest new regulations, and use fear and intimidation to prevent miners from reporting on the neglect to measure or control it as federal safety regulations dictate. By these tactics and avoiding workers' compensation payments, often through long, grinding court cases, the companies are externalizing the true costs of mining again, further jeopardizing and sacrificing the lives of their own workers by ignoring increased rates of incurable black lung disease.

Another dimension to social injustice raised by Morrice and Colagiuri (2013) relates to differentials regarding power and control, which include financial inequities. The coal companies have had more social, economic, and social capital than the miners as illustrated by their influences over national and regional resources and through their abilities to stall or oppose badly needed regulations on environmental health and safety. This is especially evident in the MTR case along with the more recent declines of social capital among the miners and their communities lack of resistance to MTR itself and to the resurgence of black lung disease. Non-miner advocates have been drawing attention to the rise in that disease. Morrice and Colaguiri (2013) cite a slurry spill in Kentucky that released 300 million gallons of toxic waste resulting in the contamination of 27,000 homes and local water supplies. The offending company was only held liable to a US$5,500 fine. The massive externalities of health destruction as documented by Epstein et al. (2011) means that the costs are borne by consumers, taxpayers, sufferers of pollution, miners, and the health care system—not the mining companies. Coal companies have too much power over financial and natural resources and unequal control over political and institutional power such as courts, legislatures, and executive branches of national and regional governments, and capacities to control information and shape information to their benefit in coal mining areas.

Coal has done too much damage to waters, landscapes, flora, fauna, and human health, spirit, and social equality. Above all it is the most significant culprit with regard to air pollution and global climate change. Of all the energy sources available to humans, coal's global reign as "King" needs to be completely abolished.

7 The "Devil's Excrement"—Petroleum

Ten years from now, twenty years from now, you will see, oil will bring us ruin. It is the devil's excrement.

Juan Pablo Perez Alfonzo in a speech in 1975
Venezuelan Diplomat and a Founder of OPEC

In the 1600s, Sir Francis Bacon promoted extractivism as a motivating ideology to discover and benefit from "nature's secrets" (Merchant 1980). Since then, one could say that its practice has not only triumphed but has run rampant. To be sure, there have been benefits. They include almost unimaginable extensions of power: to save ourselves from labor; to enhance our personal comfort; to accumulate ever-increasing amounts of material affluence; to find endless ways to entertain ourselves through electronic gadgetry; to give us the capacities to travel virtually anywhere in the world in a day; and to even explore the heavens through sophisticated rocket machinery.

Yet at what environmental costs—the destructive and costly consequences of mining and drilling for fossil fuels would probably be the first things to come to mind. Extraction for energy conjures up images of mountaintop removal, strip-mining, gushing oil wells, huge ocean oil spills from deep-water drilling, oil spreading, sinking tanker ships, and massively degraded landscapes. After damaged environments, comes the ever-widening consensus that these fuels generate polluting and greenhouse gases that could threaten most life forms with extinction. *Extraction of fossil fuels and their use have to be considered the number one global environmental issue.*

While such industries have been damaging, they have been the most profitable and powerful ones in history. The *Guardian* (https://www.theguardian.com/business/2020/feb/12/revealed-big-oil-profits-since-1990-total-nearly-2tn-bp-shell-chevron-exxon) reported that, from 1990 through to 2020, the big four petroleum companies—Exxon, Shell, Chevron, and British Petroleum—accumulated just shy of $2 trillion in clear profit while exercising the most dominating global political influence of any economic sector. They were able to escape any serious liability for the environmental damages they created. From 2008 to 2012 during a major boom of $100-plus-a-barrel prices, over $9 trillion of new oil money entered into the global economy and there was a rush from all sorts of people to cash in on this, and much corruption and outright theft occurred as Alexandra Gillies (2020) details. Yet in this context an International Monetary Fund working paper (Coady et al. 2019) estimates that governments globally annually subsidize oil companies by about $5.2 trillion through direct payments and by taking on the extraneous costs of poor health, global warming, and environmental destruction.

Petrostates

Along with staggering wealth and environmental costs, negative social consequences have also been accumulating. For developing countries, many of these circumstances can be summed up through the phenomenon of "petro-states" and the process of "petrolization" as analyzed by political scientist Terry Lynn Karl (1997, 1999) and by anthropologists Stephen Reyna and Andrea Behrends (2008) as the "crazy curse" of oil. Petrostates became significantly noticeable in 1973 and 1974 when the Organization of Petroleum Exporting Countries (OPEC) discovered its own power by charging five times the then usual prices for petroleum. This generated a crisis of oil shortage, leading to rising costs of living for just about all commodities, promoting poverty, inducing inflation for some sectors, such as the elderly, within their usually more affluent customers' societies, and restricting or reversing positive development in poor countries. All the countries in OPEC were themselves still developing, and some, such as Nigeria, had only relatively recently escaped colonialism. In the initial euphoria from newly discovered political and economic clout, their leaders felt confident that, with their newfound importance in the global economy and with oil riches from continued production, they could open fast lanes to development that would include manufacturing and GDP wealth to rival the West.

They became addicted to easy oil revenues and some exhibited a tendency to throw money at problems throughout both booms and busts. They were often rash and inefficient in making policies while obsessed with consolidating their own highly centralized power. Politicians avoided equally sharing revenues and making decisions about what to do with them with the leaders and citizens in the various regions of their countries. To do so could have reduced poverty and unemployment, and brought about effective social and health services to tangibly improve their fellow citizen's quality of life. Instead, tragically the oil-producing hinterlands tended to suffer the most, especially from environmental and health consequences, and living conditions have tended to actually deteriorate rather than improve—quite significantly in some cases.

For these oil-producing, often fragile nation-states, multiple and compounding negative consequences were prone to accumulate. For a start, they invariably experienced the "Dutch disease"—rapid and compounding inflation and costs of living due to large influxes of external capital, money that flowed in through petrodollars. This made their other products too costly for other countries to buy and thus hollowed out their agriculture systems' potential for export, and any attempts to develop manufacturing were similarly futile. This meant that only a few people had the ready cash to buy large amounts of luxury goods from abroad such as expensive cars, yachts, private jets, jewelry, computer equipment, and whiskey as well as many off-shore investments such as mansions and condominiums. These overwhelmingly went to a small elite that grew ever tighter in its self-enhancement, corruption, cronyism, nepotism, and a tendency to resort to violent solutions when their wealth and power were threatened. Complicated webs of off-shore shell companies, secretive bank accounts, and deposits in tax havens such as the British Virgin Islands became the means for these kleptocrats to safeguard their gains (Gillies 2020).

Through the over-emphasis on oil production, these elites rely on *rentier* economies where revenues come through the leasing of their country's lands to oil companies often in joint partnership with national oil companies. Historically rentier economies have been criticized by economists from both the right and left because they produce nothing new, are exploitive of others, and rely on easily gained sources of revenue requiring no diligent effort or frugality in management when abundant. Rather than taxation, such revenues can account for over 90% of petro-state budgets. In their rush to develop pet mega-projects, and in spite of much cash already flowing in, these governments sometimes borrowed huge amounts of money. When global markets fluctuated due to oil gluts and prices suddenly crashed, this hindered capacities to continue repayments at a steady rate. Even then, the oligarchies, nominal stewards of their nations' wellbeing, frequently continued to borrow even more to maintain usual standards of high spending while waiting for revenues to increase. Vicious circles of mismanagement and financial waste spiraled. Through all of this national elites are rarely accountable to voting taxpayers and, in most cases, they are governed by military juntas, single party-rule, or a few hereditary rulers such as in the case of Saudi Arabia and the Gulf States. Much of the rent collected and sometimes in the many billions of dollars can simply disappear usually into unlisted, offshore bank accounts with little or no judicial recourse.

The military in these countries, which may themselves govern, demands a large share of gross domestic product, around 10% compared to the more normal 1.5% to 3% range. It uses that to gain the most advanced weapons to suppress citizens or subdue rivals reinforcing the already authoritarian nature of these countries. Millions have been killed in civil and external wars since the 1960s when OPEC first formed. Because of the curse of oil, these countries can be the most violent, divisive, and in many cases most unstable. Causes can be reduced to greed from the potential of unimaginable and easy rentier wealth with many rivaling factions seeking to control it. In cases such as in Latin America and Africa, there has been revolving government rule by the military, authoritarian rulers or dictators, and occasional nominal democracies all competing, often violently, for the power that comes through controlling the extraction of oil.

The most disturbing social outcomes include the realities that in so many cases there have been significant reductions in per capita incomes and increases in the numbers that drop into the most impoverished statuses—the reverse of what one would logically expect. During the earliest booms, population growth soared, but later there were declines in economic growth and real standards of living, with levels of inequality and poverty deepening. Lawlessness, gangs, drug dealing, black markets in many commodities, including stolen petroleum, and environmentally destructive sabotage persist in some countries. The well-armed police and military do little to stop it and often profit from it themselves.

The oil fields themselves often exist in the hinterlands of minority, marginalized, and persecuted peoples even in countries usually not thought of as petro-states such as the United States and Canada. In Canada we can point to Cree, Dene, and Métis in Alberta's tar-sands, and to the Inupiat in Alaska's North Slope and the Beaufort Sea's massive deposits as examples of those who have faced great disruptions. In petro-states such marginal yet strategic positioning can be devastating and has led even to attempts at genocide as with example Saddam Hussein's campaigns against the Kurds in Iraq, the huge amount of poverty, murder, violence, and environmental injustice suffered by the Ogoni of Nigeria, and the dislocation of Indigenous peoples in Eastern Ecuador.

Throughout all of these violent and socially dysfunctional conditions, the transnational petroleum companies, headquartered primarily in liberal democracies, are too often freed from any ethical standards, public pressures, scrutiny of the press, and laws of the home countries that might otherwise restrain them. They are, in effect, significant accessories or even instigators for such corruption, crimes, and injustices as Michael Watts (2004a, 2004b) regarding Nigeria and Suzanne Sawyer (2004) with reference to Ecuador, and others have informed us about. Also, to sum up this petro-state and "crazy curse" phenomenon Reyna and Behrends (2008: 11) state, "the curse is the paradoxical situation where what should have brought good brings bad"—and often very bad.

The rest of the chapter will focus on the notorious tar sands operations in the northern regions of Alberta, Canada. Is Alberta, a Canadian province, really a petro-state? Is Canada as a whole? Resources are handled by the provinces in one of the world's most decentralized federations. Some of Karl's (1997, 1999) other criteria are missing such as authoritarian rule, corrupt or ineffectual courts, and close to 100% of government revenues for operations. Yet in such a debate, long-time observer and journalist specializing in the Alberta oil patch, Andrew Nikiforuk (2010), *does* declare Alberta as a petro-state—even all of Canada as such.

His opinion was based on the Stephen Harper Conservative federal government of 2006–2015 when the Alberta tar-sands dominated Canada's economy with long-term expectations of production increases five times that of the then present. Canada had ambitions to be an energy superpower based on its position of having the world's fourth largest reserves of petroleum—95% of which was in Alberta and mainly located within the tar-sands. At the time, oil and gas accounted for 25% of Canada's exports, and the government was aggressively trying to get more pipelines approved for delivery to ports with access to Asia and refineries in the American South. Harper's government suppressed its own scientists trying to report on climate change and fiercely attacked environmentalists in public relations wars. Environmental impact regulations were gutted and the biddings of transnational petroleum companies were given first priority in policy making. Alberta, Canada's provincial surrogate petro-state, was only charging 30 cents royalty on a $60.00 barrel of oil yet depended on 30% of its government's large-scale operations on oil production and without a sales tax as with all other Canadian provinces. At the same time, Canada was suffering from the Dutch disease because of an overly strong dollar for the rest of its export sector. It should be noted, though, that since this time and more recently, Alberta did suffer through a bust in its oil industry due to global surpluses and even occasional negative prices due in part to the COVID-19 crisis. Many thousands of workers lost jobs, and petroleum stocks plunged. Such volatility is reflective of the experience of petro-states.

There have, then, been reasons to maybe call Canada along with Alberta a "petro-state-like" complex but fall short of calling it a petro-state in the same way as Kerry Lynn Karl (1997, 1999) would conceive of one. The tar sands and gas fields of Canada's Mackenzie-Athabasca Basin are presented here as an extended case study because of the massive environmental damages, and the large number of Indigenous peoples that have been negatively impacted.

The Alberta Tar-Sands

After the Amazon and the Mississippi, the Mackenzie-Athabasca River drainage system is the third largest in the Western Hemisphere. It drains one-fifth of Canada—1,805,200 square kilometers—and includes a large portion of the Northwest Territories, a considerable amount of the Yukon Territory, a portion of northern Saskatchewan, and the northern half of Alberta. Compared to southern and eastern Canada, the region is low in population—less than a half million and mainly living upstream in Alberta. Fort McMurray dominates with about 70,000 people. Most settlements number in the few hundreds or several thousands. First Nations—Western Woodland Cree, Dene, Métis, and Inuvialuit—are found in larger concentrations outside of Fort McMurray, Peace River, and Yellowknife. Until the 1970s, most of this region was relatively unspoiled wilderness. But since then, development of the extraction kind threatens on a massive scale (Ervin 2011). The basin, including deposits offshore in the Beaufort Sea, contains trillions of cubic feet of natural gas. Its oil reserves, at 165 billion barrels, are fourth only after Saudi Arabia, Venezuela, and Iran (https://www.alberta.ca/oil-sands-facts-and-statistics).

The Canadian tar sands projects are collectively, the world's largest private industrial enterprise as well as its largest energy development and they have brought the country much notoriety as a major polluter and greenhouse gas emitter. The tar sands are found in an area roughly size of Florida, 142,000 square kilometers, divided in three parts in the Peace River and Athabasca Rivers and Cold Lake regions. It involves massive strip mining and clear-cutting with the world's largest earth moving machines. The initial process is energetically quite expensive involving the equivalent of one barrel of oil to produce between three and five barrels of tar-sands oil by removing the oil-containing bitumen and it requires huge amounts of water. This compares to other jurisdictions such as Texas where one-barrel equivalent will energetically process thirty or so barrels (Nikiforuk 2008).

The facilities there definitely pollute—they have been the leading causes of acid rain and the largest single-source of carbon dioxide emissions in Canada. Landscapes are blighted by thousands of square kilometers of seismic lines, industrial roads, gargantuan clear-cuts, strip mines, and tailing ponds. They were being produced at 3.3 million barrels a day in early 2024 (https://www.alberta.ca/oil-sands-facts-and-statistics).

Water environments are most vulnerable to pollution and disruption, but the long-term consequences regarding tar sands operations are barely known. They require huge amounts of water with up to 2.3 billion liters drawn from the Athabasca River each year. That is equivalent to that consumed by a city of 2,000,000—twice that of Calgary, a city of over one million for local reference. The water is used in the process to remove bitumen from tar sands and then heavily polluted it has to be kept in tailing ponds for a yet unknown time. In places where "overgrowth" is too deep, pressurized steam is sent through pipes to loosen, melt the tar sands, and then be removed by conventional drilling. That steam also becomes polluted water. Can the Athabasca River and other sources sustain this with reduced glacier melting, low winter flows, and demands of a potential three to five-fold increase in production that at one time had been projected? The toxicity and impacts of the tailing ponds are not fully understood while it continues to leak into the Athabasca River and groundwater. Yet incidents with duck flocks when they have landed on wastewater holding ponds have been quite telling. They all died within a few hours. So, any conclusions that there are likely to be serious impacts with wildlife and humans downstream on the Athabasca River are well-founded (Nikiforuk 2008).

Fort Chipewyan is on the western end of Lake Athabasca where the Athabasca and Peace Rivers form a delta before flowing into it. It is home to 1200 Dene, Cree, and Métis who are highly dependent on lake water, wildlife, and especially fish. People have complained of unusual tasting, soft-fleshed, deformed fish, and a deep unease with their drinking sources with many turning to bottled water. For such a small community, their resident doctor, John O'Connor, in 2006 publicly exposed extremely high rates of renal failure, lupus, leukemia, hyperthyroidism, and cancer. Especially significant were several cases of a rare bile duct cancer, cholangiocarcinoma not found even in populations of 100,000 (Nikiforuk 2008). A later study by Kevin Timoney (Timoney and Lee 2009) found many contaminants in the Delta including arsenic, mercury, polycyclic aromatic hydrocarbons, and that were all high and increasing compared to upriver locations. He maintained that if U.S. Environmental Protection Agency standards were used people would not be allowed to eat the fish. These findings were confirmed and dramatically expanded with photos of tumor-displaying, deformed fish in a study done by Dr. David Schindler of the University of Alberta, an internationally renowned hydrologist and biologist. Clearly, with heavy metals of cobalt, lead mercury, arsenic, and mercury entering soils and waters along with polycyclic aromatic hydrocarbons in the air, soils, and waters, the plants, animals, and waters that humans depend upon cannot help but become toxic in time. (https://thetyee.ca/News/2010/09/17/AthabascaDeformedFish/).

The Lubicon Lake Cree reside in a territory between Athabasca and Peace Rivers and claim 10,000 square kilometers as their traditional territory. They did not, however, until very recently have a land reserve recognized by the Canadian government. When Treaty 8, covering certain Indigenous groups in Alberta and some in Saskatchewan was being negotiated, the treaty commissioners in 1899 neglected to record their presence. They live in a zone noted for heavy oil and gas production, and, since 1979, their territory has been invaded by the oil industry—thousands of employees and about 2,000 oil wells. Thousands of kilometers of pipelines and cut lines disturb their territory and livelihoods with polluted fisheries and game becoming scarce. There had even been a rise in tuberculosis a bellwether indicator of poverty and social stress.

For four decades the situation of the Lubicon Cree represented a major example of First Nations social injustice with the future of their Cree culture and subsistence base at high risk for extinction. Both the Alberta Government and the Federal Government stalled in negotiations for a just treaty settlement since they were both more interested in maintaining the extraction of oil and the revenues it generated. There were blockades, boycotts, and censures of Canada's treatment of the Lubicon Lake Cree with support coming through the World Council of Churches, Amnesty International, and the United Nations. Finally in October 2018, a settlement was reached with the establishment of a 246 square-kilometer reserve near the village of Little Buffalo and $121 million in provincial and federal funding (https://www.canada.ca/en/crown-indigenous-relations-northern-affairs/news/2018/10/lubicon-lake-band-alberta-and-canada-celebrate-historic-land-claim-settlement.html).

While perhaps not as egregious in the use of naked violence as in Nigeria, or Ecuador (Watts 2004a, 2004b; Sawyer 2004), the actual environmental damage may be significantly greater than in those countries. That claim is important when considering potential and actual greenhouse gas emissions from the operations themselves and from those emitted by vehicles eventually powered as a result of these extractions. Considering that the oil itself, bitumen based, has to be considered the dirtiest of all the global deposits while being the fourth largest of the remaining reserves. Its components also make diluted bitumen much more corrosive in pipeline deliveries resulting in a much higher probability of oil spills (Huseman and Short 2012). Vast amounts of forest—thousands of acres—and meters deep of earth, creating gargantuan deep pits, have to be removed, and wildlife and fisheries are severely threatened in those zones. For instance, seismic cut-lines and more intense activity in Alberta's far north have further threatened already highly endangered woodland caribou herds.

We do not know yet the full impact upon Alberta's First Nations peoples but the predicaments of the Lubicon Lake Cree and the Fort Chipewyan Indigenous communities should give us some clues. Generally, previously, and comparatively, while residing in their boreal regions, Canada's Indigenous peoples there had more opportunities for self-reliance and sources of decent nutrition than had been the case with those of reserves in southern heavily settled regions. That circumstance has been deeply disturbed by all this recent industrial-styled massive development.

Canada as a First World nation along with its provinces has had advanced environmental impact assessment (EIA) and to a lesser extent social impact assessment (SIA) requirements for its resource industries. It has, perhaps, stricter safety requirements and ongoing monitoring and inspection procedures at pipeline, wellhead, and other petroleum operations than you might find, for instance, in Nigeria or Ecuador. Yet one gets the impression that since neoliberal policies came to dominate in the 1980s, along with the austerity-argued downsizing of departments of the environment, there has been a significant erosion of the effectiveness of these laws; that projects are fast-tracked for approval; that oil company agendas are privileged; and EIAs and SIAs are done perfunctorily. At the time of writing, and as a bizarre example of the Province of Alberta's privileging of oil interests, the government has suspended *all* environmental inspections due to the supposed impossibility of keeping to social distance protocols during the first COVID-19 wave of 2020. Yet at the same time, it permitted haircuts and was promoting its two large cities, Calgary and Edmonton, as destination cities for the resumption of the National Hockey League season (https://globalnews.ca/news/6968696/aer-suspends-monitoring-nhl/).

These suspicions of laxity and the favoring of company agendas are further reinforced if one reads the reviews of government and industry reports as analyzed by anthropologists Clinton Westman (2013) and Patricia McCormack (2020) especially as related to the SIAs relevant to First Nations peoples. Westman and McCormack also inform us that the monitoring of impacts, both environmental and especially social, over the course of oil sands operations, is poorly developed almost to the point of non-existence. Furthermore, the shallowness in understanding of socio-cultural complexities and impacts in these reports is disturbing. Unfortunately, there is a tendency among EIA and SIA consultants to provide governments and corporate sponsors with what they want to hear since their businesses rely on getting one contract after another most frequently from the same employer.

Only recently have researchers started to rectify the vital knowledge gap about the impacts of extraction in this little studied region of Canada's boreal forests. The trailblazer has been Clinton Westman (2013; Westman and Joly 2019) who has been collaborating with his graduate students as well as other mainly young researchers. A compendium *Extracting Home in the Oil Sands: Settler Colonialism and Environmental Change in Subarctic Canada* (Westman et al. 2020a) has begun the process of documenting the details that we need. First, we are reminded that this region, larger than many nation-states, has been treated as an internal colony of Alberta. While the majority of its people have been Indigenous and in the history of resource development, now intensified with petroleum extraction, there has been a clearly unequal "distribution of risks and benefits best characterized by the dynamics of colonialism and environmental racism" (Westman et al. 2020b: 4).

We can begin in 1870 when soon after achieving nationhood, Canada acquired from the Hudson's Bay Company and the British government the Northwest Territories and Rupert Land, parts of which later became the provinces of Alberta and Saskatchewan. Federal Treaty 8 commissioners in the late 1800s, backed by evidence from the Canadian Geological Survey, hastily went about exploring the regions while taking note of where bitumen oil and valuable minerals were to be found. They were able to almost completely avoid the placing of reserves for First Nations on locations having such resources that First Nations could themselves commercially develop. Supposedly, in the treaty-making process, the "duty to consult" was recognized as the means for First Nations to cooperate and even co-manage if Euro-Canadians proposed development in these regions or to register opposition. Yet the "duty to consult," a weak instrument for Indigenous people, has never, as a veto, led to the cancellation of a project in Alberta, only in reality to the notification of the proposed project and what impact it might have and perhaps some modifications (Huseman and Short 2012).

Indigenous peoples were generally allowed to continue hunting, collecting, and fishing on their traditional territories, subject, though, to possible legal modification if Euro-Canadian governments saw fit. Later, as part of a general process to accommodate white trappers during the Great Depression, First Nations families were also permitted to register trap lines outside of their reserves. To Cree, Dene, and Métis, these became exceptionally important for not only managing their resources in the then important fur trade but for family harvesting of moose, deer, caribou, fish, geese, ducks and their eggs, medicinal plants, berries, and other plants in the organization of a still relatively autonomous subsistence. While most no longer trap, these small areas are still vitally important for families to make a living and to maintain their culture in the actual practice of it. Yet these trap-lines are legally considered the sources of economic "profit" and the rights to them can be abrogated if the Province wishes to approve different types of commercial development especially petroleum on those small but vital parcels. Such broad arrangements, however, allowed northern Albertan Indigenous peoples to remain relatively secure in subsistence up until recently (Huseman and Short 2012; Longley 2020). Overall, it is clear, nonetheless, that the whole treaty-making process had been stacked in favor of Euro-Canadian settler interests and as per usual the Indigenous peoples were never made fully aware of settler intents and motives.

There has been some small progress in rectifying some injustices, mainly around the inadequate size of reserves. Canadian federal court cases from the 1960s onward have upheld the legitimacy of particular First Nations land claims. In relevant cases, the appropriate amount of land was not designated in the original negotiations, such as for Treaty 8, in determining the final sizes of reserves. A process known as treaty land entitlement was instituted and meant to provide additional land to those First Nations deprived of the proper amount—usually caused by oversights in tallying all band members at the time of making a treaty. It did, however, require careful and often protracted research and negotiations among the First Nation, the federal government, and the province in question to determine and select remedial amounts in per capita acreages. The process was complicated because much of the traditional land might have already been turned over to white settlers as private, deeded property as was certainly the case in the southern farming parts of Saskatchewan and Alberta.

The Mikisew Cree First Nation, in the far northeastern part of Alberta, was able to have such a claim ratified and selected 42,000 acres in Wood Buffalo National Park, a federal property. In addition, it tried to choose about 55,000 acres more that it was entitled to well within its traditional territory that included lands rich in tar sands bitumen. This claim was logical as to use and occupancy and would have provided the potential for resource revenues to benefit impoverished people. Longley (2020) documents how the Government of Alberta stalled and frustrated the whole process for 14 years in order to wear down the Cree's resolve, and who eventually abandoned that option so that settlement could be finally achieved in 1986. Again, this underscores the Settler authority's colonial process of dispossessing indigenous peoples land base in order to increase its own wealth.

Land and waters have not only been essential for the maintenance of subsistence vital to a people but it has been and is the base for the continuing reproduction of the culture itself. Janelle Baker (2020) gives us a small but fascinating taste of that as part of her research monitoring the impact of oil sands activity on wild food supplies. She focuses on berry patches and their central role in human subsistence but also on bears that are prominent in spiritual beliefs and intimately tied to humans in connection with such berry patches. Stories are essential for First Nations enculturation processes with children and reinforcing existing beliefs among adults. What is important to underscore is *that this living culture is taught in the context of actual practice on the land*—in berry patches, while hunting and rendering wildlife, at fish camps, and on the trap line. Baker recounts some tales about bears and their important relations to humans and to other creatures and the land.

Controlled fires were a traditional subsistence technique for First Nations people to manage a landscape creating corridors for ease of travel, to generate or renew hay meadows, and to create berry patches. Fires were set to cultivate pastures for deer and other grazing animals to eat, thus making them available for hunting, and the berry patches of blueberries, cranberries, huckleberries, and others were a major part of the diet for people who had no domesticated plants. The controlled burning of moderate-sized patches served significant ecological purposes of renewal—creating circumstances where new fresh plant shoots could emerge, attracting grazing animals and releasing nutrients from the ash into the soil. They also reinforced biodiversity by making travel easier for animals and plant exchanges. These burnings, since banned with the establishment of Euro-Canadian jurisdiction, also served to prevent or minimize the occurrence of the large-scale and destructive forest fires that happen much more frequently today—but part of Nature's cycles nonetheless.

Fire is considered as a life-giving force—in one origin story, the Creator breathes fire into two poplar trees to make human beings. Fire can be considered a life-renewing and cleansing force making land fresh and being referred to as "grandfather fire." Humans came as a later part of creation, being preceded by other animals that were able to acquire greater wisdom in relating with all of nature and its forces. Humans need to learn from and respect such animals. Bears are extremely important in this context being considered in many ways similar to humans, understanding the Cree language, and frequently referred to as "grandmother bear" in their encounters. Like moose and deer, bears can allow themselves to be killed to be eaten by humans but under certain circumstances can also retaliate in kind for offenses. Bears are highly dependent for their survival, especially before hibernation on berry patches. Humans while also placing major importance on berry patches should recognize that priority be given to bears—it is their food first and foremost and the bears are sharing it with humans. More of these stories about bears were told to Baker while berry picking and checking out the degree of contamination of them from petroleum activities as part of her applied investigations with Cree co-researchers. They revealed the subtle teachings as would be taught to children in this and other activities on the land and on trap lines as essential ways of reproducing Cree culture—in effect the land was the classroom. It also shows a complex, spiritually grounded, ecologically aware, and sensitive living culture. As one indicator of their awareness of ongoing change she recounts how the Cree are worried, because of the reduction of berry patches due to industrial contamination, that bears are being forced to rely more on human garbage dumps leading to contamination of both the bears' and their own sources of food (Baker 2020).

Another article in the Westman et al. (2020a) collection by Lena Gross (2020) recounts an interview with a Métis oil worker who, to his deep regret, followed orders by the foreman to clear a patch of bush by using diesel fuel during the winter when it was hard to start fires. He knew it to be a bountiful berry patch that now would be severely contaminated and not grow berries anymore. The Métis felt morally upset by the act but explained that he and his family needed to work so that they could eat. Gross's article raises the issues of those who work in these oil sands fields—both Indigenous and non-Indigenous. The huge industrial transformations in the region generate a situation where it becomes necessary or even attractive for local peoples to take jobs within the industry. Significant players such as Syncrude can accurately boast that it is the largest employer of Native people in Canada and wages can be quite high—a potential benefit for chronically impoverished people in an economy overwhelmed by capitalism and the pressures of money and wage labor. Yet, at the same time this creates a strange condition whereby they are becoming a rural proletariat in their very own homeland now dependent upon foreign, transnational oil companies.

Gross concentrates on the role and high incidence of accidents both regarding workers' injuries and those frequently imposed upon the environment in the form of oil spills and major polluting leakages. These create significant risks for humans and wildlife, and to conceptualize it all she borrows the construct *wastelanding* from Tracy Voyles (2015). The term was originally used in context of uranium mining on the Navajo reservation in the American Southwest to refer to settler, government, and corporate disregard for a land still held sacred. Through the attitude of environmental racism, Navajo land was seen as one that could be sacrificed as a wasteland of destruction and pollution for the benefit of a larger good and was justified, even legally, as being "natural."

Gross (2020) applies that notion to realm of tar sands not only to the spills, leakages, clear-cutting, and gigantic earth gouging and removals but also expands it to the *"wastelanding" of human bodies* through the accidents and working conditions of the workers—both Native and non-Native. Wastelanding is an attitude and process of violent disrespect toward Nature and people alike. Besides the long-term pollution on human health, it includes wear and tear from the high rate of industrial accidents that is associated with the challenging working conditions can involve long shift work hours in extreme subarctic cold and often in the dark far from any community. In turn that often contributes to behaviors both individual and social that add damage to bodies, society, and nature—all influenced by the desire to push production for high profit and wealth.

In keeping with these observations, a former pipeline worker once told me of frequent and relatively standard operations. Pipelines transporting diluted bitumen require frequent cleaning through reaming with wire brushes—upon hitting a corner where a turn is made, the reamer may easily burst through a corroded pipe and thus create an oil leak. Workers tend to be in a hurry to get their jobs done not only because of company demands but, for example, because of the extreme discomfort of working in deep sub-zero conditions in dark, freezing, subarctic February nights. The corroded and broken pipe may be eventually fixed, but the oil leaks are often not reported as they are legally required due to the paperwork, the mandatory fines, and the extra time-consuming and costly overtime work to clean up oil spills. Dr. Kevin Timoney (2017 https://www.nationalobserver.com/2017/02/02/news/researcher-says-alberta-regulator-underestimating-impact-thousands-oil-spills) published a report investigating over 60,000 oil and saline water pipeline leakages in Alberta over a 37-year period and concluded that the Alberta Regulator was neglecting such unreported spills as well as significantly overestimating the success of clean-ups, much to the continuing damage to Alberta's north.

In their review of existing EIA and SIA literature, Westman and Joly (2019) write about some of the known socioeconomic impacts. Oil companies will make agreements with particular First Nations providing employment and community facilities, but these are confidential and they appear in some cases to involve hundreds of millions of dollars. These acts of generosity seem to be most frequent in the context of the First Nation removing its particular objection to the company's proposed enterprise in their traditional territories. In some cases, as especially exemplified by the Ft. McKay Cree First Nation just north of Ft. McMurray with full employment, ready cash reserves, there have been lucrative co-ventures with oil companies. Furthermore, Ft. McKay with some tar sand deposits on its own territory was considering developing its own mine operations. Yet the reserve is already surrounded within 30 kilometers by seven major oil sands mines. A study in 2016 by Alberta Health showed heavy air pollution likely causing severe health problems with 13 air-borne chemicals measuring above odor and environmental thresholds. These include hydrogen sulfide, benzene, and sulfur dioxide, all known to be extremely dangerous (https://edmontonjournal.com/business/energy/study-finds-air-problems-around-oilsands-community-fort-mckay) with benzene being a major carcinogen. The Ft. McKay First Nation might be a forerunner example of a people formerly rooted on the land making the transition to a fully industrial life. Could it turn into a new "cancer alley"? What will be the long-term social consequences also remains to be seen.

The costs for cleaning up Alberta's oil operations have been estimated at $260 Billion CDN and that is likely to be proven an underestimate in the end as is usually the case with such estimates about environmental damage. Furthermore, the Pembina Institute (2017) has calculated that, after 50 years of operations, less than 0.1% of the tar-sands operations have undergone the environmental remediation that is required by law and at the expense of the oil companies (https://www.pembina.org/blog/fifty-years-oilsands-equals-only-01-land-reclaimed#:~:text=Since%201967%2C%20of%20the%20940%20sq%20km%20disturbed

,and%20returned%20to%20the%20province%20%28See%20Figure%202%2). Given the likelihood of the hundreds of years still required for any such reclamation to be completed, and if actually done, the burden will most likely be shifted to the Canadian taxpayer. We have precedents already with the Gunnar Uranium Mine in Saskatchewan (https://thestarphoenix.com/business/local-business/gunnar-cleanup-to-exceed-250m-10-times-estimate), Sydney Tar Ponds in Nova Scotia (https cleanup-1://www.cbc.ca/news/canada/sydney-tar-ponds-get-400-million-.488265), and gold mines in the Yukon and Northwest Territories costing the Canadian government billions of dollars after the companies disappeared or went into receivership. There is little reason to believe that the oil sands will be any different but likely to be much more expensive given the very much larger size of its operations in comparison to these precedents.

Conclusions

We have *all* benefitted to some extent from the use of fossil fuels in making our lives convenient in so many ways. The questions here have been centered on impacts upon the peoples residing in the areas where the raw commodities are extracted and especially what are the impacts upon those who are indigenous to those territories. Beyond the frankly dirty and polluting nature of such industries, do the locals benefit overall or do they suffer more on the whole?

Regarding Alberta, distinguished historian and policy analyst Ken Coates (2016) takes a much more sanguine point of view than I do. While conceding that the presence of a strong environmental anti-pipeline, anti-oil sands, Indigenous movements has been publicly quite dramatic, he suggests that First Nations peoples are far from unanimous in their opposition. He points to important court cases and federal and provincial policy orientations that have emphasized reconciliation with Canada's original inhabitants. Negotiations with the parties involved have led to supposedly significant and lucrative business partnerships or particular agreements between particular First Nations and oil companies, such as the Ft. McKay Cree with a 49% investment, jointly with Suncor Oil Company, in a $503 million, oil holding, storage operation. First Nations leaders based on their own acumen plus legal and policy changes are displaying ever-increasing agency and capacity to shape development in their own regions and this should be a strategy for all of Canada's First Nations people in becoming full partners in Confederation seems to be Coates's central message.

Parallel to Coates, but through a detailed and critical political-economic analysis, Ian Urquhart (2018) also notes the widespread participation of First Nations employees in tar sands development as well as a leadership willingly engaged in joint projects. He suggests, however, that they really had no choice. With the rapidity of the boom beginning in the 1990s, they were presented with a "fait accompli" and "their leaders more often than not made the rational choice: accommodation" (Urquhart 2018: 9). He argues that free market fundamentalism, the driving ideology of neoliberalism, overwhelmed *all* dimensions of the massive growth of what will likely be North America's biggest-ever resource boom. Between 1999 and 2013, over $200 billion (CDN) was invested in its expansion and 2014 set a new record with $34 billion.

Free market enthusiasts claim that unfettered markets, by themselves without government interference, will always lead to the most rational and socially beneficial results. Yet the tar sands enterprises were aided and abetted by the Alberta and federal governments at every turn. To begin with an egregiously low royalty rate of 1% was placed on gross returns of petroleum sales, whereas one as high as 50% had previously charged on net earnings with other petroleum projects. This only encouraged companies to massively keep expanding the scale of their operations leading, among other things, to almost obscene waste and likely to an eventual glut of stranded capital. Environmental regulations were tailor-made to suit industry desires, for instance, making it impossible for third-party organizations, such as Greenpeace or the Sierra Club, to raise objections about the environmental consequences of particular projects. At one point a large and valuable wetland region notable for its large and unique avian nesting and breeding ground was considered for a provincial wildlife preserve status, but tar sands lobbyists were able to get the decision vetoed. Their rationale seems Orwellian in that they argued that this would lead to the "sterilization" of any tar-sands deposits found in the region that would be forever lost to extraction. Any kind of policy intended to reduce greenhouse gas emissions, for which the tar sands are notorious, was essentially a form of greenwashing. Petroleum companies often paid $15 a ton for their carbon releases, rather than attempt any technical solution for carbon reductions simply because they were so flush in cash (Urquhart 2018).

Considering First Nations again, we could speculate on a scenario where they might have chosen, in a concerted way, to try to block tar-sands development on the basis of traditional rights and pollution and at the time of the expansions. They would have found themselves vastly outmatched and out organized by the enormous legal and financial resources of both government and the corporations. While the circumstances of Alberta do not match Terry Lynn Karl's (1997, 1999) definition of a classic petro-state such as Ecuador or Nigeria, perhaps Urquhart's (2018) case study suggests another set of criteria for the concept. Alberta certainly looks like a jurisdiction that completely privileges petroleum interests even in ways that damage its own environment and underserves its own people by collecting revenues today and for future generations.

Recognizing the reality that many First Nations people *do* wish to (or have to) participate in these developments, Clinton Westman's research with his graduate students concentrates on issues of consultation. The *duty to consult* is recognized in virtually every treaty and legal decision surrounding Canada's relations with its First Nations. However, the meanings and the practice of that phrase have remained nebulous. Ideally it would assume an equal partnership with regard to decision-making in the first place. Then it should involve ongoing communication and joint action regarding land and water use management throughout any activity carried out by government or corporations on traditional lands. That could involve the placement of wells, mining pits, pipelines, roads, incidents of blasting activities, oil spills, flooding or draining of wetlands, lakes, and rivers, placement and design of tailing ponds, and setting of fires. Supposedly government departments and corporations have officials, protocols, and agreements to do just that. Similarly, First Nations governments frequently have personnel in their band offices to engage in liaison with their government and corporate counterparts. To date this level of consultation is abysmally underdeveloped, perfunctory when present, and with First Nations more often than not left in the dark rather than full partners in consultation (Westman and Joly 2019).

While First Nations leaders appear to have been forced by contingencies into a limited choice of co-optation or remain *both* polluted *and* poor, not all did so. One striking example is Chief Alan Adam of the Denesuline First Nation of Ft. Chipewyan on Lake Athabaska, where health and environmental issues were highlighted earlier in this chapter. He was the subject of a *New York Times* interview (https://www.nytimes.com/2020/06/28/world/canada/allan-adam-indigenous-alberta.html) due to his being subjected to well-publicized incident of egregious police brutality in Ft. McMurray. Adam was elected chief in 2007 and, because of the aforementioned health and environmental damages, immediately instigated a lawsuit seeking the rescinding of all the oil leases that had just been granted in Dene traditional territory. His community lost, but since then he has been part of over a dozen similar lawsuits and has become an international symbol of resistance. However, in 2018, even he signed an agreement providing certain accommodations for employment and community development with Teck Resources that had proposed the largest tar sands mining operation yet. His explanation was that he was tired and exhausted from all the failed lawsuits and wanted his people to benefit in some ways similar to other First Nations jurisdictions. Ironically, the Teck project was later withdrawn because of market uncertainties. It is worth taking into account his last comment in the interview, "Our people will be environmental refugees, because the land will be polluted so badly by the oil industry. I figure we've got 25 years left" (Chief Allan Adam, https://www.nytimes.com/2020/06/28/world/canada/allan-adam-indigenous-alberta.html).

Like Coates, it would be inappropriate for me to make any generalizations regarding all of Alberta's Native peoples. Yet I will also make note of just some considerations from their environmentalist side here. During two 3-day periods in the summers of 2013 and 2014, I attended the last two Athabasca Tar Sand Healing Walks in which First Nations elders led over 500 Native and non-native activists including environmental luminaries Winona LaDuke, Bill McKibben, and Naomi Klein. Several days of Indigenous ceremonies, workshops, and talks were held in the encampment before the actual walk. Those talks, including ones by Chief Allan Adam, focused on the particular sufferings that specific First Nations were experiencing due to the petroleum industry invasion of their lands. The highlight was the 14 kilometer walk that followed the route of the gigantic Syncrude Oil Company's operations with the most dramatic earth removing, tailing ponds, and refinery operations in the tar sands region—those closest to Ft. McMurray with the facilities most photographed in coverage of the operations.

There were two things I found especially notable. One was the lack of hostility to us among the many workers we encountered—in fact many honked their horns or showed other signs of support. To me that indicated a possible underlying recognition that all is not right with this kind of extraction, and perhaps that some might feel restricted by their own opportunities or skills and would not mind alternative forms of employment. The second notable observation was the deep and persisting spirituality and articulation of a sense of moral crisis in this massive threat to their lands in the speeches of the elders and of the remarkable, new, university-educated cadre of young activists. Highly articulate and rooted in Indigenous identity and beyond their particular First Nation, they have formed networks and alliances with 350.org, Keepers of the Athabasca, Greenpeace, and the North America-wide Indigenous Environment Network and have been organizing and speaking at national and international rallies

Environmentalists, settlers or not, owe First Nations activists much gratitude for their front-line action in resisting oil development in fragile environmental zones, and because of greenhouse gases and global warming. These are issues that face us all—including the grandchildren of oil executives.

8 Hydroelectric and Irrigation Dams

Hydroelectric dams appear benign because they produce electricity using renewable energy and, unlike coal and petroleum, they do not emit carbon dioxide when generating power. They offer major enhancements to agriculture since green revolution techniques require massive amounts of irrigation water. This is particularly so when farming may be ramped up to two or three crops a year as in parts of Asia. Besides being a mechanism for managing flood risks and sometimes enabling more effective river transportation, they can represent potential sources of clean water for people living in regions of scarce and polluted water. Development specialists such as economists and engineers with less holistic perspectives on costs and benefits than anthropologists and ecologists can be quite enthusiastic about them.

Yet dams do have worrisome social and environmental consequences and among the environmental costs is that they too can be emitters of greenhouse gases through rotting vegetation that causes methane releases from flooded lands. Most serious are matters of extreme social injustice. By building them, governments have displaced millions of their own, often most vulnerable, citizens and deprived them of their land and water-based livelihoods with little to no compensation. By 2000 the World Commission on Dams (2000) estimated that from the end of World War II up to the beginning of this new millennium large dams had globally dislocated 40–80 million people. Furthermore, Michael Cernea (1997) calculated that approximately four million people are similarly removed *each year* by dam construction. So, since over 20 years have passed since the Commission's report, it is possible that upwards of 100 million have been involuntarily removed by now. The original higher calculation—80 million—may be more likely, in my opinion, since it is to the advantage of countries to underestimate the numbers thus reducing any liability costs.

Calling for research and advocacy, *involuntary resettlement* looms large in applied anthropology as it does in political ecology (see Oliver Smith ed. 2009a; de Wit ed. 2005; Cernea ed. 1991; Hansen and Oliver-Smith ed. 1982). Large dams have epitomized social and environmental injustice—it is hard to think of more extreme examples other than slavery or outright genocide. People quite understandably tend to resist any attempts to oust them from places where they have traditionally lived and where they are committed to time-tested and familiar

DOI: 10.4324/9781003495673-10

ways of living—the obvious practical or utilitarian dimensions of any person's life. Also consider certain powerful emotions that are hard for most people to express yet have vivid personal meanings attached to them. They tend to be deeply felt personal identities of people seeing themselves as inseparable from particular places. People with such sentiments have specific, localized, hard-wired memories of formative experiences that were first forged in childhood and then consistently reinforced throughout their lives. They, their kin, and neighbors through unique cultural ideologies and practices may perceive these places as sacred—as their "holy lands" (Stoffle and Evans 1990)—further making them, along with other beings such as plants, animals, and spirits, inextricable from very *specific places*. When there is a removal from these connections, the results can be catastrophic in ways that we, who are not the victims, will never completely understand. Remove a people, kill a culture is a tragedy alltooften discovered by anthropologists studying cultural change. These important intangibles are impossible even for the most sympathetic observers to accurately portray—let alone the fact that they are overlooked, ignored, or dismissed as trivial by technocratic, development-enthusiastic, urban-raised, upper-middle class, and top-down planners who advocated the building of the dams in the first place. They are not the ones to suffer.

The disruptions and displacements for dam constructions have proven to be destructive economically, socially, and culturally for the victims of such relocations a majority of the time. In some cases, the results can be considered within the realm of ethnocide if not genocide (see Johnston 2009 for the horrendous case of the Chixoy Dam in Guatemala). Involuntarily displaced people have rarely (perhaps never) been satisfactorily compensated or rehabilitated for their losses. And as citizens of their nation-states, surely, they should, if anything, benefit from their countries' development and not bear the majority of the costs. Because of the work of anthropologists, geographers, and sociologists engaged in social impact assessment, we have come to a better, but far-from-complete, understanding of the negative consequences of dam building as the issues of *development forced displacement* have received more attention.

Environmental, Health, and Social Impacts of Major Dam Construction

Anthropologist Thayer Scudder (2005) is the world's most experienced researcher on the impacts of hydroelectric and irrigation dams having also been a commissioner on the World Commission on Dams (2000) initiated by the World Bank and World Conservation Union in 1997. His research goes all the way back to the 1950s when he investigated the impacts of the Kariba Dam in Zambia where 57,000 Tonga people were left impoverished and today still not successfully adjusted to their long-ago forced relocation. With his (2005) writing and while still critical, he thought of dam building as a "flawed option," reluctantly necessary for development in impoverished nations. Yet later (Scudder 2019) he had come to the conclusion that they were definitely much more socially and environmentally damaging in the long term than they were worth. He had become especially disillusioned with the activities and policies of the World Bank—the major promoter and lender for such development projects as building dams. It had not effectively implemented the recommendations of the World Commission on Dams on how to resolve the huge social and environmental impacts that the Commission had revealed. Furthermore, the Bank greatly reduced its social impact research into displacement issues caused by its funding of such projects. He states that large dams are "a major component of a dysfunctional international paradigm" elaborating:

> Five all too common major reasons for unacceptable resettlement outcomes are lack of political will, lack of implementation capacity, inadequate finance, inadequate resettler participation, and lack of viable opportunities for resettlers to improve their household and community livelihood.
>
> (Scudder 2019: 17)

As he pointed out in his 2005 *The Future of Large Dams: Dealing with Environmental and Political Costs,* they are not just simply touted but have been doggedly pushed for by politicians and development coalitions—by the World Bank and regional development banks, large construction firms, engineers, contractors, economists, politicians, and bureaucrats representing urban and middle-class interests. Dams have been advocated as the fast track to modernity most especially in Asia, Latin America, and Africa. Dams, again though, are accompanied by much suffering, though, in the form of displacement of large numbers of people—the term *development forced displacement* (DFD) has been coined to underscore the coercion exercised by the state (Oliver-Smith 2009b). All too frequently the ones displaced are those already impoverished, rural, and marginalized such as tribal peoples, peasants, and Dalits or what used to be labeled "Untouchables" in India. Those benefitting tend to represent urban and middle classes as well as rich farmers who can afford the benefits of any resulting irrigation systems—let alone the benefits to corporations, often foreign, such as mining and construction companies and vested interests within nation-state bureaucratic fiefdoms. Such inequities show the frequent and overwhelming injustice to the practice.

Scudder has been researching what has been designated as large dams—ones that block major rivers for the purposes of generating electricity for powering large and growing cities hungry for electricity and for irrigation projects that will supposedly enhance the agricultural capacities of their regions. They are classified as large when they are at least 15 meters high or between 5 and 15 meters and have storage capacities of three million square meters or more. Many are much larger than these minimums—by the time of the World Commission on Dams more than 50,000 had been built since World War II and, as of 2021, there are almost 60,000 of them. Beyond that, there are millions of smaller dams since there are over 2,000,000 in the United States alone (Moran et al. 2018). It is only with the large dams that decent records are kept but, even then, there are appalling gaps of information. This is most especially with regard to the numbers actually displaced and details on their conditions of rehabilitation after displacement.

One hundred and forty countries have built such dams—most notably earlier in the United States, Canada, among European nations, and more recently in developing countries such as Brazil, Turkey, India, and China. Cost overruns are standard—96% over the projected costs on average (Ansar et al. 2014), and there are many cases of financial corruption and faulty, inadequate construction standards overlooked to reduce costs in the context of that corruption that could lead to later environmental disasters (Scudder 2019).

Already there has been more frequent flooding associated with them—as a result of the heavier rains now associated with climate change that will become further exacerbated over time. In other places, long-term droughts becoming more frequent can greatly reduce the amount of water in the reservoirs seriously jeopardizing the tasks for which they were built and the natural flowing integrities of the rivers from which they were conceived. This has been seriously dramatized with the disruption and over-drawing from the Colorado River (USA), the major drying-up of its reservoir (Lake Mead), and the jeopardization of the hydroelectric capacity of the Hoover Dam itself—the original standard for such development (Moran et al. 2018). Large water reservoirs often associated with the dams could eventually complicate local weather patterns in ways that we cannot yet predict as climate change amplifies. Tied to the holding back of water is the standard phenomenon of silt build-up in the reservoirs due to water being slowed down before reaching its ultimate destination. Over time the silt built-ups will overcome much of the storage ability created by the reservoir and add to the possibility of more frequent flooding in areas when and where rainfall has increased (Moran et al. 2018).

Then there are large releases of natural gas resulting from decaying vegetation, notably in areas that were previously forested but now flooded over by the reservoir (Deemer et al. 2016). Annually and collectively, it contributes about 104 million metric tonnes of methane to the atmosphere (Orrego 2012). Methane has at least twenty times the warming impact of carbon dioxide, and this represents about 15% of anthropogenic greenhouse gas emissions. Added to this in the form of carbon dioxide are all the activities involved in their construction, most especially when, as with many mega dams, concrete is a major part of their foundation. Concrete production accounts for 8% of carbon dioxide releases annually. Not all dams are constructed from concrete, but most are, and given the millions of them having been already built along with their sizes, that represents a lot of concrete and carbon dioxide.

Downriver impacts are considerable with water flows from dams often restricted to trickles or even virtually absent in some cases, leading to significant habitat fragmentation. Wild and domesticated animals suffer from less water; some species of fish and other riverine species disappear or are greatly reduced, sometimes through their spawning territories drying up. Groundwater and aquifer recharging is diminished. Riverine estuaries, where before the construction of dams, sediments, both organic and non-organic, flowing down-river provide essential nutrients for spawning fish and other ocean wildlife. Tiny diatoms, for example, which are at the basis of the aquatic photosynthetic food chain, require the mineral silica that would normally be brought down to an estuary (Orrego 2012). Many more environmental consequences can be found within complicated riverine ecologies.

Human health impacts are significant especially as related to the large bodies of water controlling flow—the lake-sized reservoirs and the slow-moving canals that bring irrigation water to distant farms. Public health officials first became aware of this very serious detriment due to the great extension of the range of schistosomiasis, a devastating waterborne fluke disease, because of the construction of the Aswan High Dam on the Nile River in Egypt (Heyneman 1971, 1979). Tropical and subtropical regions, where a good portion of contemporary dam building is done, are especially vulnerable to diseases whose ranges are vastly expanded through water. These include schistosomiasis, malaria, encephalitis, hemorrhagic fevers, gastroenteritis, intestinal parasites, and filariasis after the dam and irrigation projects (Lerer and Scudder 1999). This is all reinforced by still and stagnant waters backed up in reservoirs, thus creating ideal zones for snail and insect vectors, and disease-carrying mosquitos can thrive in weeds along the lake-sized reservoirs.

People's nutrition may be hampered by the loss of highly fertile soils that were previously at the edges of river sites now extensively flooded, along with demands and food insecurities of the now higher population densities due to reduced available land. Accidents during construction and later flooding can cause serious casualties. Displaced populations can suffer from inadequate water and generally polluted and unsanitary conditions. Food insecurity is often made worse among those displaced. Stress-related conditions can be quite profound for those now forced to make major social and economic adaptations while losing their treasured traditional home places. These can include cardiovascular conditions, diabetes, drug and alcohol abuse, and suicides (Kedia 2009). Also mentioned by Lerer and Scudder (1999) are conditions of poor or polluted access to water sources both up and down rivers and diseases affecting cattle—for example, liver flukes and trypanosomiasis.

The negative social and economic impacts of hydroelectric dams constitute what are probably the most alarming. An important earlier summary of them was by the biggest promoter of such dams the World Bank itself through its then-senior advisor on social policy—Michael Cernea (1997). He pointed out three scenarios of impact: boomtowns during construction downstream communities, and those populations subject to displacement and resettlement. Of the three, the last category was the most serious and damaging. With boomtowns during construction, large numbers of imported workers devoted to construction can disrupt a local region in many ways especially if it is based on traditional kin and primarily subsistence orientations. Among them could be peaks in teenage pregnancies as a result of casual affairs with imported workers, school drop-outs from those seeking construction jobs, the possibilities of increased substance abuse, violence, disease transmission, frequent road accidents, and a disrupted boom-bust, cash-dependent economy after the project is finished.

The second scenario, involving downstream communities, involves primarily disruptive socio-economic consequences. Fishing can be severely damaged because of the reduced access of fish to spawning grounds or in numbers to those restricted flow waters. In areas where there had been wide low valleys, the residents might have been beneficially dependent for their agriculture on annual flood waters that brought rich soils with abundant organic matter—but no longer. Scudder (2005) reports that the World Commission on Dams calculated that those downstream experiencing negative impacts are probably 10 times in number to those in the dam site regions, including those displaced. That would mean that, so far at least 500 million plus are thus affected—so, not an insignificant category by any means.

The impacts on displaced peoples undergoing involuntary resettlement after their homelands have been flooded are again the most serious and lasting. They include homelessness, joblessness, food insecurity, landlessness, marginalization, loss of access to communal property, and community disarticulation, plus deterioration in health. The important social ties of people's networks, including most especially kin ones, that are lost and had reinforced mutual aid and psychological support can bring about serious marginalization and lowered status in the new communities where they are frequently now strangers. The amount of suffering and unhappiness experienced can be extreme. Important symbolic and even sacred sites, such as places of prayer and ceremony and ancestor burial grounds of immense emotional importance may have to be abandoned. The exacerbation of existing poverty can be amplified and the regional economy is thrown way off balance because of the damaged social links and destroyed institutions (Cernea 1997). Indian anthropologist Singh (2020) adds to this list with economic risk categories such as the loss of traditional skills, loss of livestock and their grazing grounds, and migrations of absent menfolk looking for work because of unemployment in resettlement areas. Sociocultural risks may include loss of cultural identity and feelings of "placeness." Psychological risks include substance abuse, increased violence against women, and child neglect.

Cernea (1997) next looked at the sorry state of mitigation or compensation for those who have been involuntarily resettled after being displaced. Much of that initially has to do with poor planning in anticipating the total costs for dam projects in all their facets. It is common to significantly underestimate the physical costs of construction. That is carried over in the anticipation of the human resettlement costs but here the expenditures tend to remain the same as budgeted. Yet funds are always raised to finish the project because that is considered the central mandate so there is no option other than to find the funds (and as we have seen they go up on an average of 96% of the project). Also, rarely are the budgets able to accurately estimate the per capita or family needs for full resettlement and key issues such as underestimating the fair prices for the acquisition of land and the actual existence of available replacement lands elsewhere. The biggest concern, almost to the point of travesty, has been the lack of consultation and ongoing participation in planning with those who will bear most of the costs and the least of the benefits—the peoples being displaced. These and many other problems such as accurately estimating the numbers to be resettled and compensated end up, in effect, *externalizing many of the costs of the dam projects upon the most victimized—the displaced people.*

Conclusions

It would seem hydroelectric power is a dubious option as *the* global mainstay for renewable, noncarbon emitting sources of electric power. It is, in many ways, another example of the recklessness in our rushed attempts to impose industrialism upon nature. Even some policymakers, have come to the realization that the practice of forced displacement can cause extreme suffering for already vulnerable and marginalized peoples.

There have even been reductions in its extent and practice in many developed countries. After the historic initiation of big dams, notably the Hoover Dam, the Tennessee Valley Authority, and along the Columbia and Missouri Rivers, during the 1930s when hydro provided 40% of American electricity, it has been reduced to 6.1% and replaced by other sources such as coal, natural gas, renewables, and nuclear power. Since the 1960s regarding the United States, dams have been considered too costly to build, with all the best sites taken, and major environmental damages including the blocking of important fish migrations. Many hundreds of dams are being removed at great expense but at much less cost than would be required to refurbish them. Even the U.S. Army Corps of Engineers has abandoned its zeal and now concentrates on other projects such as protections from coastal erosions. While still significant in countries such as France, Norway, Sweden, and Switzerland with high-altitude sites, the same trend is occurring in Europe with almost 3,500 dams removed (Moran et al. 2018).

It is in the developing world where many people are still without connections to electrical grids that hydro dam production has been ramped up. This boom is, however, by no means, restricted to the benefits of public consumption. In many cases, they are for the convenience of foreign mining companies as in the Congo or in Central America and South America. Targeted rivers and peoples undergoing or threatened with displacement can be found in major river basins such as the Amazon, Congo, and Mekong. The last with many millions of people contains a highly productive freshwater fishery with a high diversity in Southeast Asia. Well over a hundred large projects with much investment from China are being built or planned with over 100,000 megawatts of electricity potential projected but with a devastated fishery feared, let alone a yet unknown number of displaced peoples (Moran et al. 2018).

As citizens of their nation-states, local peoples really should benefit from any such projects and should participate in the planning and be consulted on the socio-cultural, environmental, and economic consequences. Yet research after research has shown that the very opposite occurs—repression, violent targeting including assassinations of those in opposition, forced displacement, worsened poverty, failed resettlement, and broken promises and neglect have been all too frequent. There is good reason to be concerned about the consequences of hydroelectric dam building booms in big parts of the developing world.

At least in passing, it should be noted that my own country, Canada, is a major producer as well as an exporter to the United States of hydroelectric power. Globally it ranks fourth in production and more than 50% of its electricity comes from hydro. What is different is that very little of the dam and reservoir combinations are linked with or devoted primarily to irrigation and agricultural development. Furthermore, the vast majority of dam sites are found far away from populous regions and instead located in the northern regions of the provinces thus primarily in the boreal forests and Canadian Shield regions. That does not mean that they are without controversy or negative impact—in fact a whole chapter could be written about the political ecology of dams in Canada. Many of the same environmental issues as previously discussed here would apply. The biggest issues are the controversial relations with Indigenous or First Nations peoples, who in these northern regions are still strongly dependent upon local resources from hunting and fishing. While not being physically displaced in any significant numbers as compared to India, they rightly complain about broken treaties that, among other things, contained clauses framed as "duties to consult," and because of mass disruptions through the construction processes, significantly reduced access to hunting, fishing, and trapping sites, flooding of sacred and ancestral burial sites, and the general disturbances of pristine wild rivers. Two projects that have received especially negative criticisms for impact are the massive James Bay Project in Quebec (Hornig ed. 1999) and Site C along the Peace River in northern British Columbia (https://items.ssrc.org/just-environments/first-nations-and-hydropower-the-case-of-british-columbias-site-c-dam-project/).

Returning to global considerations, unlike coal as described in Chapter 8, however, there is, at the same time, no urgent reason to completely abolish hydroelectricity as a power option. Existing hydropower plants that are well-designed and relatively benign in their environmental and social impacts can supply a significant part of a nation's electric needs under favorable circumstances. Although there is always at least some environmental cost, that is also true of wind and solar power—although significantly much less so in scale and extent. Dams can even serve as forms of energy storage in the form of pumped hydro since so far no massive storage battery system has been made technically feasible. Unused photovoltaic and wind power that would be otherwise wasted or not available when needed can be stored. The excess energy from the latter two can be used to pump water upward behind dams and then later released through generators as hydropower when the sun is not shining and the wind is not blowing. Norway is highly significant in its direct hydropower and in developing such storage capacities. Its hydro produces over 33,000 megawatts because of the advantages of its terrain and natural water flows meeting as much as 98% of its own electrical needs, trading much of its capacity to the rest of Europe. It is notable, especially for a renewable partnership with Denmark, a wind power leader, and Norway might eventually serve as touted as "Europe's Battery" (https://www.hydropower.org/region-profiles/europe). Also considering wider uses, small hydroelectric, or "run-of-the river," projects that do not involve the blockage of any river flows are also very promising in a future alternative regime that is highly variable. We will return to these ideas in Chapter 12.

9 Uranium and Nuclear Power
The Case Against

It is clear from histories of nuclear energy that its development was subordinate and linked to military agendas right from the beginning and, in many respects, still through to the present (see Rhodes 2018). Enrico Fermi accomplished the first, successful, nuclear-fission, chain reaction at a makeshift reactor at the University of Chicago on December 2, 1942. This was the essential precondition for the Manhattan Project to develop an atomic bomb before the Germans could. All things nuclear became highly secretive and as Rhodes (2018) puts it "totalitarian." These two characteristics of the industry persist to today along with a knowledge base that is far from public understanding. This generates the need for an inordinate amount of trust in highly specialized expert opinion and a need to be ever alert to the possibility of vested interest.

A number of steps beyond the military monopoly led to the emergence of nuclear power as a source for electrical energy. One was the establishment in 1946 of the U.S. Atomic Energy Commission (USAEC), which was extremely powerful and secretive in its operations, but whose commissioners were civilians. It maintained ownership over all fissionable materials in the United States and tried to purchase as much as possible from other sources including Canada, which had provided both uranium and directly processed plutonium for American nuclear weapons (Harding 2007). After the Soviet Union had detonated an atomic bomb in 1949 and with Dwight Eisenhower's coming to the U.S. presidency in 1952, the Cold War went into feverish overdrive with a massive nuclear arms race.

However, to offset the image of the United States as a warmonger with countries that were being courted for alliance, Eisenhower in 1953 introduced the "Atoms for Peace" Program to the UN General Assembly meeting. He offered to extend American knowledge, uranium, and other fissionable materials to an international atomic energy commission to act to improve agriculture, medicine, and with special attention to global energy needs in a rapidly expanding age of electricity. Several other countries including Canada, Britain, France, and Russia had already started to research electricity-producing nuclear reactors, and the field became quite competitive. Russia was the first to be successful in 1954 with a five Megawatt reactor, and that spurred the American effort even more.

The next step was accomplished under the supervision of Admiral Hyman Rickover, who had been the leading developer of a nuclear navy. In 1953, Rickover was doubly assigned to the Navy and USAEC where he was put in charge of a program working with the Westinghouse Electric and the Duquesne Light Company in the building of the Shippingport Power Plant. Beginning in 1957, its reactor provided electricity for the Pittsburgh area.

By the 1960s, American, Canadian, British, and French interests were all rapidly expanding nuclear power plants and entering international competition for sales of their reactor designs. The United States promoted privately owned facilities such as through Westinghouse and General Electric now both actively engaged yet with all aspects still ultimately controlled by the USAEC. In other countries such as Canada and France, state-owned corporations designed reactors although they sometimes sold the designs to private energy companies. The Soviet Union expanded its program and also built reactors in its Eastern European satellites (Rhodes 2018). In that Cold War era, depending on the country, nuclear reactors could still be used to continue to support national weapons programs by generating plutonium or enriched uranium 235. This sometimes occurred inadvertently from the perspective of the uranium seller. During the 1970s, Canada outlawed the use of its uranium and its nuclear reactors for nuclear weapon development. Nonetheless, India then Pakistan, after purchasing a new generation of CANDU heavy-water cooled reactors from Canada, both did exactly that—i.e., developed their nuclear weapons programs by means of Canadian technology and uranium (Harding 2007).

The 1960s and 1970s experienced a boom in nuclear power extension with slightly over 400 reactors being built with a quarter of them in the United States. By the 1980s 16–17% of the world's electricity was provided by nuclear sources. There have been various waxings and wanings of the industry due largely to the major crises of meltdowns at Three Mile Island, Chernobyl, and Fukushima, massive cost overruns in building reactors, and on-going lack of solutions to contend with the piles of nuclear waste that have been globally accumulating at nuclear sites. Nevertheless, proponents have been promoting the notion of a "Nuclear Renaissance" as the solution to save humanity from global climate change. Approximately 100 reactors of various designs are being planned in the Near East and Asia, mainly China and India (World Nuclear Association 2019, https://www.world-nuclear.org/information-library/current-and-future-generation/outline-history-of-nuclear-energy.aspx). Its enthusiasts also advance nuclear power to be the way of expanding developing economies in a world ever hungry for more energy.

I oppose these options. The debate, though, is essential, and I will return to it during the conclusions of this chapter as well in later chapters on solutions. The rest here will be an examination of the circumstances of uranium mining, nuclear power, and the possible location of nuclear waste disposal in Canada concentrated in my home province of Saskatchewan. A case study later especially draws attention to environmental movements, and, in this case, I was an inside observer and participant. It was remarkably successful—even beyond our own expectations.

Uranium and Nuclear Power in Canada

During the period 1930–1940, radium, then the world's most valuable mineral, was mined at the eastern end of Great Bear Lake in the Northwest Territories. In 1942, Eldorado, a crown corporation reopened the mine to extract uranium, which is often found accompanying radium. The U.S. government ordered 850 tons, which were refined at Port Hope, Ontario, for the Manhattan Project. Two hundred and twenty tons directly supplied the project and was mixed with uranium from Colorado and the Congo. Canadian uranium was thus used in the atomic bombs that were dropped on Hiroshima and Nagasaki in 1945.

First Nations men, Sahtugot'tine Dene ("Slavey") from the community of Deline at the western end of the Lake, were employed to transport the ore in burlap bags on boats that crossed Great Bear Lake where it was put on barges going south on the Mackenzie River. The men were not warned of the risks of radioactivity. El Dorado did not provide them with protective clothing and they were not advised to shower. Deline later became known as the "village of widows." Although no precise epidemiological figures are available from this remote community, 14 of the 30 former Eldorado employees died early of cancer. Villagers lamented the loss of able hunters and future elders to teach oral traditions. The Village Council later sent a message of apology to Japan since at the time they did not know what the uranium was for. The Canadian Federal Government has denied any responsibility for the unusual rates of cancer. In 2002, it did, however, agree to clean up the abandoned mine with the toxic waste, and tailings still leaking into Lake (http://www.ccnro.rg/dene.html).

By 1955 the mining of uranium shifted to Elliot Lake, Ontario, where its products served the American nuclear weapons program until 1962. After that it supplied uranium for the Canadian reactors largely in Ontario but by the 1990s shut down because of the rapid market ascendancy of Saskatchewan's mining. Elliot Lake has had a chequered history with regard to occupational health with high rates of lung cancer among miners and radioactive pollution in Anishinabek First Nations Territory (Harding 2007).

In 1942 Canada started early among the very first countries in the nuclear industry with a joint Canada-United Kingdom experimental project in Montreal, sharing information with the American Manhattan Project. They designed a heavy water reactor that was built at Chalk River in the Ottawa River Valley in 1944. In 1952 the Atomic Energy of Canada Corporation Limited (AECL), a crown corporation, was created to design reactors for university and government research, produce medical isotopes, and design electrical power facilities. Canada, although an advanced nuclear power, formally forsook any ambitions to build nuclear weapons, but that did not exclude Eldorado (the government-owned mining corporation) from selling uranium to Canada's allies for their weapons programs.

Beginning in 1961, AECL started designing and building a series of its CANDU reactors in Ontario, Quebec, and New Brunswick. Quebec would later shut down its sole reactor, and the one in New Brunswick has been plagued with operational problems, being offline for four years (2008–2012) at great expense. Eighteen reactors operate in Ontario, providing about 60% of its electricity, with nuclear energy providing about 16% of Canada's total electric production.

It is interesting to note that Bruce Power, the private operator of eight of the reactors in Ontario, employs a SWAT team armed with automatic rifles to protect its reactors in case of terrorist attacks that attempt to spread radioactivity among civilian populations—a need that would obviously be unimaginable in the case of a wind turbine or solar panel power station. When Ontario Hydro was privatized in 1999, it was found that $15 billion of its $20 billion debt was due to its nuclear operations and that excess has been passed on to consumers through surcharges on their electric bills to pay off this public debt.

AECL also sold CANDU reactors to Argentina, South Korea, India, Pakistan, Romania, and China. Again, the sales to India and Pakistan unfortunately contributed to both India and Pakistan (bitter enemies to each other) developing an atomic arsenal. Because of that fiasco, Canada is now very careful to avoid selling uranium and nuclear technology for military purposes (http://www.ccnr.org/india_pak_coop.html).

In 2011 a large portion of AECL's operations was privatized with the commercial development of nuclear reactors leased to the Montreal-based, engineering giant SNC-Lavalin. That action is questionable since that company had been labeled one of the most corrupt transnational corporations in the world. The World Bank blacklisted it for any of its projects in 2013 (https://www.worldbank.org/en/news/press-release/2013/04/17/world-bank-debars-snc-lavalin-inc-and-its-affiliates-for-ten-years). Most of Canada's nuclear reactors have reached or surpassed their recommended time limits and will have to be refurbished or decommissioned soon at great expense. The majority of these operations will be under the control of SNC-Lavalin.

Uranium in Saskatchewan

Here I recount a case study (Ervin 2012) and one that illustrates most of the issues raised about uranium mining and nuclear power. It also reveals the remarkable work of a set of environmental movements that show the capacity for citizen resistance. This is important because one of the central messages of this book is that in order to resist the very many environmental wrongdoings of this current era, it requires the diligent and determined efforts of social movements. They are also required to establish the principles and actions to reverse the damaging trends of excessive extractivism and ever-aggressive and profit hungry neoliberalism. Both of the latter are the root causes of global environmental crises.

Saskatchewan, where I live, is the world's largest producer of high-grade uranium ore—that having Uranium oxide (U3O8) close to 15% of the ore content. Kazakhstan surpassed Canada as the overall largest producer in 2009, but up until then Canada, mainly through Saskatchewan, had provided about 20% of the world's uranium. This uranium is found in the geological formation known as the Athabasca Basin in the far northwestern part of Saskatchewan that is also associated and adjacent to the Athabasca tar sands region of Alberta, thus of high strategic interest to the extractive industries. The most dominant companies have been Orano the French national company and, above all, Cameco the largest, publicly traded, uranium company in the world headquartered in Saskatoon. There have been other companies exploring and engaged in mining but Orano and Cameco dominate supplying France's large fleet of reactors in the case of Orano and domestic and global markets in the case of Cameco. The largest markets for Cameco have included the United States, China, Japan, Canada's reactors, and Europe. As a transnational corporation, Cameco has operations in Australia, Kazakhstan, and the United States and it manages milling and processing of the fuel in various sites in Ontario. Its sales operations in Europe have been an issue of controversy—by creating a nominally separate company with minimal staffing in Zog Switzerland that has a purposely and notoriously low corporate tax rate. Yet selling uranium produced in Canada the parent company has been able to avoid since the 1990s over $2 billion CDN in federal and provincial taxes while directly delivering to European customers. Revenue Canada had been in a protracted court battle to retrieve such presumed tax losses while Cameco claims such deferrals represent good and legal business practices on behalf of its shareholders—it lost and Cameco prevailed at the Supreme Court level basically because it was assumed that corporations seeking mechanisms that are yet to be declared illegal is normal behavior irrespective of any presumed moral consideration (https://www.bloomberg.com/news/articles/2021-02-18/cameco-finally-puts-to-rest-13-year-long-tax-dispute-with-canada#xj4y7vzkg).

Through the mid-1970s most of the uranium mining in the province was done close to the town of Uranium City on Lake Athabasca by a series of companies that preceded Cameco but are no longer in existence. Among the abandoned mines was the Gunnar operation, having served the United States nuclear weapons program, it is the source of much ground and water contamination from the leaching of its tailings and there is concern about the levels of radioactivity in the lake itself which is already under threat because of tar sands contamination from the western or Alberta end. The costs for cleaning up the Gunnar Mine have been estimated at over $250,000,000 CDN and both the Province of Saskatchewan and the Canadian government are in dispute as to which should bear the costs (https://www.cbc.ca/news/canada/saskatoon/uranium-gunnar-mine-cleanup-cost-saskatchewan-1

.4114674). There has never been a thorough investigation into the environmental as well as social and human health impacts of the industry in the province. The mining conducted today is done at fly-in camps not at settlements such as Uranium City.

In 2008 a new political party, the Saskatchewan Party, came to power. Almost immediately the new premier expressed the view that Saskatchewan should become the "Saudi Arabia of Uranium" and set that agenda in motion by appointing in the fall of that year a special committee called the "Uranium Development Partnership" (UDP) whose task was to explore the options available. Significant and powerful players beyond the government itself were enthusiastically in support of expanding the uranium and nuclear industries. Such stakeholders included the mining companies, the major newspapers, TV and radio networks, media pundits, the Provincial as well as local chambers of commerce, implicitly the two university administrations, and a majority of public opinion according to polls although one could claim that the public was not fully informed on all the issues surrounding uranium and nuclear power. The broad notion of value-added uranium and nuclear development was appealing for economic reasons. Saskatchewan mined uranium but did not get any employment, tax, or GDP benefits beyond basic extraction and milling to yellowcake. Refinement and processing of fuel was done in Ontario at plants owned by Cameco. Boosters saw Saskatchewan becoming an energy power to rival more populous and prosperous oil-rich Alberta. Saskatchewan already had some major energy assets with the second largest oil production in Canada, uranium and coal mining, but the ambition was enhanced with the notion of uranium refineries and nuclear power plants along with major research centers favoring nuclear power at the two universities. Nuclear power plants could supply "carbon-free" energy for Saskatchewan's future and in its surplus provide lucrative energy exports to the United States. Another extremely attractive feature was that surplus energy could be directed to the Athabasca oil sands developments in northern Alberta. As mentioned earlier, the oil sands refining is inefficient in that it takes one-barrel equivalent of fossil fuel equivalent to produce three to five barrels of oil compared to the more standard one to thirty ratio thus exacerbating the CO_2 release dilemma. Electricity produced from carbon free nuclear reactors in nearby Saskatchewan would supposedly solve that problem was the industry's touting.

Because of the expectation that the Saskatchewan government would enter into nuclear development and for the first time legislate and financially underwrite the building of large nuclear power plants, Bruce Power of Ontario became involved. Bruce Power was the largest manager of nuclear power plants in Canada owning eight reactors in Ontario including the largest reactor complex in the world on Lake Huron. It started to look at and even negotiate with landowners in Western Saskatchewan close to the Alberta border along the North Saskatchewan River since nuclear reactors require massive amounts of water for cooling. The company also looked at sites in Alberta because it was interested in resolving the energy equation in refining the heavy bitumen drawn from mining the tar sands.

With great fanfare in March 2009, the Saskatchewan government released the report prepared by the Uranium Development Partnership's Report (2009). There were five major recommendations in that report with multiple adjunct suggestions with each general one. The major recommendation was focused on the desirability of building nuclear reactors in Saskatchewan. The UDP suggested the rapid approval of two or three large-sized nuclear reactors to a total of 3,000 Megawatts that would meet 45% of Saskatchewan's future needs as well as the presumed export opportunities listed earlier. Another highly controversial recommendation was focused on the high desirability of situating Canada's deep "high level spent-fuel repository" somewhere in the Canadian Shield of Northern Saskatchewan. At the time, 45,000 metric tonnes of fuel had accumulated in various stages of cool or dry storage at the 19 nuclear power reactors in Eastern Canada. It was widely recognized that this amount and any future quantity had to be stored in some underground safe, geologically stable facility at some point because of the radioactive dangers of the spent fuel that would last for tens of thousands of years. The suggestion was for Northern Saskatchewan since it was the location of the Precambrian Shield that many presumed to be the most stable geological formation in the world. Thus, any tunneling done there for a repository was presumed to be safe from any future geological events, flooding, and the possibility of internal heat from the metal waste containers creating cracked leakages in the rocks. There was also a presumed ethical argument for placing the high-level spent fuel in a Saskatchewan site because that is where it originally came from and the province had supposedly profited from its mining, sales, employment, and taxes. There was a pay-off bonus as well since they claimed that the construction of the project would introduce $40 Billion into the provincial economy.

Another recommendation concerned training and research. It called for a major research and development facility at one of the two provincial universities. Such a facility would contain a small-scale reactor for continuing research and potentially produce medical isotopes for the diagnosis and treatment of cancer. Facilities at the university would provide training for those involved in mining and the operation of reactors. This complex was meant to establish and maintain Saskatchewan's intended position as the global leader in nuclear issues as part of the premier's aspiration to make it the "Saudi Arabia of uranium" (Uranium Development Partnership 2009).

Soon after the release of the UDP report, a series of four stakeholder meetings was held—two each in the major cities of Regina and Saskatoon to present and promote the findings. This was all done in the context of what was considered a start to a "Nuclear Renaissance" in the early 2000s. The dangers of global warming as a result of the excessive releases of carbon dioxide in fossil fuel were more in the public scrutiny and such nuclear enthusiasts and those with stakes in the industry as obviously the members of the UDP had touted nuclear power as a "carbon-free" path to the future. To some extent the horrors of the near meltdown at the Three-Mile nuclear plant in Pennsylvania and the actual meltdown at Chernobyl had faded from public consciousness. New licensing permits were being released in nations where there had been an unofficial moratorium on nuclear power plant construction. It did look as if the nuclear industry was about to go through a rebirth of sorts. There was an expectation that the Saskatchewan government, already proven as an enthusiastic supporter of all things nuclear would go ahead and legislate with financial support the UDP findings. Yet in the spring of 2009, the Government announced that it was establishing a one-person commission named the "Future of Uranium" to consult with the public about the UDP recommendations before proceeding. The commission was to be led by a respected recently retired most senior civil servant, Dan Perrins, and to be held in thirteen communities throughout all of the regions of the province including once again, the two major cities of Regina and Saskatoon.

Movements of Resistance

There had previously been some environmental and social justice movements against uranium mining and early proposals for nuclear power and uranium refining in the province. Anticipating the possible presence of the Nuclear Renaissance campaign, a new organization was formed by less than ten members in 2006. I was one of them. The organization was called A Coalition for a Clean Green Saskatchewan (CCGS) with the objective of organizing resistance. Yet instead of mere opposition, we decided to promote non-carbon emitting alternatives for energy sources—wind, solar, biomass, and run-of-the-river (non-dammed) hydroelectric. Saskatchewan had among the very best natural resources and climatic conditions for such alternatives.

The rapid news-making events of the spring of 2009 also led to the rapid formation of other resistance groups including those in towns that were possibly targeted as sites for nuclear reactors. Since we in Saskatoon were at that point the most organized and the city was most central for all the activists, now numbering in the hundreds, an organizing and strategizing meeting was held there in March with about 60 attending. Besides the new-formed community nuclear resistance organizations, representatives of the Saskatchewan Environmental Society, a local of the Sierra Club of Canada, social justice wings of mainstream Protestant and Catholic churches, plus a number of people not associated with any particular organization. We all had previous activist experience and were united against nuclear and uranium expansion and recognized the urgency of our commitment due to the UDP report and the upcoming Future of Uranium commission hearings to start soon. Collectively all groups took on the existing label—A Coalition for a Clean Green Saskatchewan (CCGS).

A non-binding steering committee was formed as well as other committees to meet digitally. I, for example, was put on a committee for drafting press releases and one for generating information about renewable energy alternatives. We agreed to share all information among those we could trust in our Coalition email listserves. It was important to keep everybody up to date as the Future of Uranium Commission was holding meetings in various regions of Saskatchewan. One important task was to inform our local populations of our positions and educate them about what we thought was important pertaining to nuclear issues. Above all, it was important to get the public to attend the crucial public forums.

The Future of Uranium Hearings

The aforementioned retired senior civil servant led the hearings of the one-person commission in May and June 2009, to be written up in August and presented in September, with the Government responding in December as to how it would implement the findings from both the UDP and the Future of Uranium. Thirteen meetings were held in towns and cities widely distributed throughout the province, including the very far North in the midst of the Athabasca Basin, the source of uranium mining. Altogether, 2,637 people attended the 13 town hall meetings, and a little more than half completed an 8-page questionnaire workbook. The meetings were quite lively, taking on the tone of passion, sometimes resembling a revival meeting. From my perspective, the positions held and presented were well-formed and presented by the participants.

The Arguments For and Against Uranium Development

Those representing the pro-nuclear, pro-UDP side broadly speaking had in summary these points to make:

- We already mine and sell uranium so why not do more?
- Constructing nuclear plants, a used fuel deep repository, and increased mining would all have beneficial economic multiplier effects benefitting Saskatchewan.
- Saskatchewan is in a strategic position to become a world leader in this economy and be at the very forefront of a nuclear renaissance.
- Nuclear power will provide our growing energy needs for the future.
- It ensures base-load capacity in energy while getting rid of coal-burning plants.
- In that regard, it will be a major solution in eliminating CO_2 global warming emissions. Nuclear power provides a solution to global warming.
- The likely excess of energy produced provides lucrative export opportunities, especially to energy-hungry United States.

126   *Energy*

- In Northern Saskatchewan, expanded uranium mining will provide more jobs and development opportunities for impoverished First Nations and Métis people.
- The technology is much safer now, and nuclear waste is minimal compared to toxic waste from other energy sources.
- We must not miss out on such economic opportunities when they arise. The Saskatchewan Chamber of Commerce in its testimony claimed that $13 Billion (CDN) was lost to the local economy when a uranium refinery near the town of Warman in the 1990s was rejected.

Those opposing nuclear power, uranium mining, and the UDP recommendations broadly made these arguments:

- Morally it is the case that nuclear energy is ultimately tied to the military. Canada itself participated in the Manhattan Project during World War II that produced the first atomic weapons, among other things providing a large amount of processed uranium from mines in the Northwest Territories and processed at Point Hope, Ontario. This persisted through the Cold War period and through the inadvertent supplying of both the Indian and Pakistani nuclear weapons programs, and as the primary source for lethal depleted uranium coating for armaments in contemporary military operations such as in Iraq, Afghanistan, and Kosovo.
- Enormous public subsidies have already been given to the nuclear industry with little benefit to the public except huge public expenditures and contributions to the national debt.
- Always in nuclear developments, there have been enormous costs and overruns, delays, breakdowns, and the eventual decommissioning costs for power plants that cost in the many billions of dollars ultimately again at taxpayer and electrical ratepayers' expense.
- The costs of nuclear power per kilowatt-hour are more expensive than any alternatives, including wind and solar. They are getting much higher as time goes on.
- Companies that advise on financial planning never recommend investing in nuclear power.
- Private insurance companies never insure nuclear projects. That always has to be done by the government.
- In all aspects of the nuclear chain, from mining to the production of electricity, there are releases of radiation that generate health and environmental dangers.
- High level nuclear waste is extremely dangerous even in small quantities—no country has found a provable solution for its disposal, and it is extremely dangerous for thousands of years.
- The nuclear industry is a failed 1950s industry when its proponents claimed that it would produce "energy too cheap to meter."

- The then current claim of a Nuclear Renaissance is a myth. The proportion and amount of nuclear energy is actually decreasing.
- We have many lessons about the dangers of nuclear power plants, from Hiroshima through Three Mile Island and Chernobyl and many other smaller, less-publicized accidents at nuclear plants.
- Nuclear power is not "clean and green" having a heavy carbon footprint at all the stages of the nuclear chain from uranium mining through the building of reactors and the disposal of waste. There are carcinogenic results all the way through the chain as well. Soon we will run out of high-grade ore and the search for and mining of lower-grade sources could be environmentally expensive.
- Nuclear power provides much uncertainty for future generations. They will be left with enormous financial debts and the decommissioning costs of reactors no longer in service. They would also be stuck with managing the toxic, carcinogenic wastes while not benefitting from the energy itself. In turn, they would have lost out on the opportunity to invest scarce capital in renewable energy.
- Instead of nuclear power it is time to look to the future and invest in much cleaner renewable energy sources such as wind, sun, geothermal, biogas and biomass, co-generation, and other sources that are truly clean and green. Saskatchewan has the right conditions to follow examples from leaders in renewable energy such as Denmark, Germany, and Spain.

Events after the Future of Uranium Hearings—Blows to the Nuclear Industry

The commissioner was to present a report on "what he had heard" in September. In the meantime, a number of crucial events that had an impact on policy occurred and they were setbacks to the nuclear industry and indirectly strengthened CCGS's argument. The Government of Ontario, which had already invested heavily in nuclear energy, suddenly canceled a $28 Billion (CDN) contract with Atomic Energy of Canada (AECL) to build two new reactors because of the higher than expected costs and the likelihood that, as usual the actual cost would be considerably more than the pre-construction estimate. The design of those reactors was a new-generation CANDU reactor that was presumed to be the same as proposed by Bruce Power for Saskatchewan. Around the same time, Ontario created an energy act that strongly reinforced alternative, renewable energy. AECL suffered another extreme blow that summer, one that turned out to be a major international embarrassment. It had produced approximately half the world's medical isotopes at its 52-year-old reactor at Chalk River, Ontario. Embarrassed, the Federal government asked for bids on alternative facilities for producing medical isotopes.

Next the attention was drawn to the issue of nuclear waste, remembering that the UDP had made a major recommendation that Saskatchewan northern Canadian Shield be the prime location for storing the waste. The Canadian

Nuclear Waste Management Organization (NWMO) is an industry group charged with finding that site. It was looking at about a dozen potential locations in Ontario as well as several in Saskatchewan that would fit the appropriate geological conditions and have a willing host community. In early September, NWMO planned a secret meeting with only invited stakeholders at a downtown Saskatoon hotel. We at the Coalition heard of it and held a prolonged demonstration in front of the hotel with clever and humorous skits and educational materials. We pointed out the dangers of leakages, especially into groundwater and aquifers with any repository, and the high risks of road accidents involving the thousands of trucks that would have to transport the fuel almost 3,000 kilometers from their sources in Ontario and even up to 4,000 kilometers from Quebec and New Brunswick. All of these trucks would pass through heavily populated areas on their way North into the Shield country, and Saskatoon, as a hub, would be especially vulnerable to such accidents.

Release of the Perrins, Future of Uranium, Report and Its Aftermath

The commissioner released his report in September. The findings were a complete surprise to us and represented an unexpected victory. He reported that what he heard from the citizens in his hearings all across Saskatchewan was that 84% were opposed to nuclear power and the situating of a nuclear waste repository in Saskatchewan. Majorities were also opposed to increased mining and the building of a nuclear research reactor on a university campus. He proposed that there be a similar research commission to the Uranium Development Partnership—one that focused on renewable, alternative energy to provide, as it were, equal coverage to a significant energy option for Saskatchewan—something that we at a Coalition for a Clean Green Saskatchewan had been advocating at the hearings. He also suggested that the Government should move very carefully and with caution toward any possibility toward nuclear expansions (Perrins 2009).

These results, far beyond any expectations that we had, obviously placed the Government in an awkward position after its public desires to transform Saskatchewan into a global nuclear and uranium powerhouse. A value-added, nuclear industry did not appear to interest the Saskatchewan public as much as it had thought, and there was clearly a vocal element that opposed all nuclear. During that time period, Bruce Power seemed to have abandoned the scene, and in the corporate media, there were sometimes statements by politicians that perhaps nuclear power was much too expensive for Saskatchewan to consider. Then finally in December, the Provincial Government announced that it was not going to enter into nuclear energy development at this time, although it might consider doing so in the future. It would follow recommendations to facilitate and expand uranium exploration and mining and work to facilitate a university research facility. It never provided a position on a nuclear waste facility yet. As we will see a bit later, that has been dealt with by other means. By most measurements then, one would have to conclude that the Coalition for a Clean Green Saskatchewan was markedly and unexpectedly successful in its objectives during what appeared to us as a crisis time.

What Accounted for the Movement's (CCGS) Success?

Given that one of the essential purposes in writing this book is to encourage people to participate in environmental movements in the spirit of political ecology as action-based, I feel that it is important to reflect upon what can make such movements more successful because with global climate change very much more grassroots tenacity is needed.

Urgency in the local context is a good place to start. In the abstract, ideas about nuclear power and a value-added uranium industry had been floated but not acted upon before. Yet in 2008–2009, we had a new government that was hell-bent on rushing ahead with nuclear power development. Encouraged by that rhetoric, the nuclear power giant, Bruce Power, actively searched for reactor sites along North Saskatchewan. Because of overly aggressive tactics, local ranchers, farmers, and townspeople in three affected communities responded and organized NIMBY (not-in-my-backyard) that were passionate and reached a wide following. Networked calls among associates all over the province for some kind of alliance were desperate, so the already rudimentary Coalition for a Clean Green Saskatchewan was called into action along with its existing anti-nuclear pro-renewable educational materials. The three face-to-face organizing meetings in Saskatoon were important, but beyond that, we relied on Internet, email, and Facebook connections to maintain momentum. Since we were so widely scattered, small in number, over a territory as large as Texas, these digital media were the key to our success, especially for sharing enormous quantities of information about all the topics that were vital to our campaign, which was essentially one of information and public persuasion. People without access to large public or university libraries were able to inform themselves through ample Internet facilities. Those on isolated ranches or farms could prepare sophisticated, knowledgeable presentations for the hearings. Facebook, even with its many flaws and temptations for wasting time, served as an excellent means for drumming up support from the first-at-non-active and then keeping them informed. Daily, updates on the subject were provided over as many as seven email listserves with specific information or with assigned group tasks. Tied to that was the fact that, even though the corporate media supported the UDP recommendations, the working press loves a "David and Goliath" story, especially one that was bottom-up and passionate. So, we received plenty of free publicity over the four months that this was the province's main news story.

Because of the high degree of urgency and dedication with a specific timeline for the Future of Uranium report and the government decisions, we as volunteers had to work very hard within that time frame. In Saskatoon with the original

CCGS membership expanded about three-fold; we met once every week, and without Roberts Rules, we were able to generate effective meetings where key decisions were made with consensus more or less each time. *Everybody had equal voice in a completely bottom-up, emergent structure—so that reinforced solidarity and commitment.*

Analyzed altogether, this movement and its campaigns were organic and emergent—they were not pre-planned, overly organized, and centralized in momentum. Individuals and networks in communities and across communities took on roles and functions as they were needed and let go of them when they were not. Energies were not wasted; personal egos were abandoned in a necessary context for maximum solidarity. People focused on what needed to be done because of the collective sense of urgency. Highly networked communication and action at the ground level is what accomplished the most in a theater of action where communication, the sharing of knowledge, and successfully framing anti-nuke pro-renewable were the most essential actions to be taken. The networks were multiple and overlapping, similar to plants that rely on rhizomes for spreading and reproduction. Using the notion of rhizomes as a metaphor, and borrowing from Deleuze and Guattari (1987) I suggested (Ervin 2012) that is what made the movement so successful and ready to emerge again should the need arise.

The Problem of Nuclear Waste

Spent nuclear fuel is the most hazardous of all the wastes produced by humans. It can threaten human health and all life forms for hundreds of thousands of years. Although its waste is unimaginably long-lasting, nuclear power has only existed for 80 years now, and globally about 450 commercial reactors have been in existence. Wastes have been kept temporarily at those sites, stored first in cooling pools of water because they are extremely hot as well as being highly dangerous since the waste still contains about 95% of its ultimate radioactivity. After about 10 years of cooling, they are transferred to metal casks that are usually kept within concrete-lined holding areas.

While being placed on the surface of these facilities, it is the consensus that more permanent solutions will have to be found because the radioactivity lasts much longer than any containers devised by humans. The most common and logical solution has been to place it all deep underground for approximately 400,000 years. The site for a repository needs to be geologically stable—for instance, needing to be able to withstand possible future geological events such as glaciation, volcanoes, tectonic shifts, and other possibilities. The site should be far from highly populated areas just in case some unforeseen events lead to major radiation releases, but even so there has to be some willing human communities nearby with a workforce to maintain the site. There has also been the optimistic hope, especially among those who have high faith in technocratic solutions for just about any problem, of eventually devising "breeder reactors" that could use some of the remaining reactivity; so, there should be some way of retrieving that which was put underground. With that in mind such waste dumps are given the euphemism "high level nuclear spent fuel repositories" with openings to the surface so the materials could be retrieved. One such element, plutonium, has been considered as possibly providing the fuel for power plants eventually replacing uranium 235. Yet plutonium is many thousand times more reactive than Uranium 235 and is always a potential terrorist threat, especially in consideration of making atomic weapons. The United Kingdom has wasted billions of pounds of public money on that possibility of plutonium being a future reactor fuel at its massive nuclear complex at Sellafield and is now stuck with a massive clean-up bill of over £260 billion because of the huge danger its failed plutonium retrieval scheme presents (https://www.theguardian.com/environment/2022/sep/23/uk-nuclear-waste-cleanup-decommissioning-power-stations).

Below as much as several hundred meters, the prototypical repository design has tunnels connecting the stored waste in canisters somewhat distant to each other since if radioactive material goes "critical" in a reaction that generates unimaginable amounts of heat, that could possibly even melt the underground formations or at the least create cracks in rocks through which radiation could escape. There is also the serious concern about ground and/or surface water leaking in and corroding canisters and circulating radiation through interconnected water cycles. A great deal of research and expense has gone into investigating such possibilities—but it is all a major, near impossible challenge since such projects and the length of time involved are unprecedented for any kind of human endeavor. So far none of these civilian "repositories" has been built or in operation, although one is being constructed in Finland.

The Search for a Nuclear "Repository" in Canada, Focusing on Saskatchewan

As recalled, the Uranium Development Project in 2009 recommended that Saskatchewan be the location of Canada's long-range, high-level nuclear waste repository. The final decision about a location is actually out of a provincial government's jurisdiction, although it can promote the idea as agreeable, or contrarily, legislate against the possibility of a location in its territory as the Province of Manitoba has done.

Ultimately determining the location is a federal responsibility and to be based on a decision recommended by the Nuclear Waste Management Organization (NMWO). It had been examining 11 places—three in Saskatchewan and the rest in Ontario—all in the Precambrian Canadian Shield, a massive rock formation perceived as the world's most stable geological configuration. The criteria for the location include the appropriate formations to store the materials safely for 400,000 years and a willing nearby host community that could benefit with employment but also has a long-term responsibility. Canada has had 21 major nuclear reactors (including major research facilities) all in Eastern Canada. As of 2019 that total of used fuel amounts to 57,000 metric tonnes (Global News May 19, 2019, https://globalnews.ca/news/5329835/canadas-nuclear-waste-to-be-buried-in-deep-underground-repository/) and the cost of building the project is estimated to be $23 billion CDN.

If such a site were to have been built in Saskatchewan, that would mean transporting the materials from as far away as 4,000 kilometers, considering that one of the power plants is in New Brunswick. It would involve truckloads of waste in metal kegs. To what extent would the truckloads have to remain relatively light because heavily packed waste might go "critical" in terms of nuclear reaction does not seem to have been addressed in the literature. It would require at least thousands of heavily guarded truckloads across most of Canada and passing through Saskatoon, a hub city of 300,000. The likelihood of one or more dangerous road accidents or an attack by terrorists along the way would remain a statistical possibility, even probability. Then there are the considerations of safety and the possibilities of leakages at the ultimate storage site itself.

NMWO had identified three potential areas in northern Saskatchewan Canadian Shield country where it was conducting geological testing and negotiating with possible willing host committees. There were two communities in the northwest of the province—a First Nations reserve (English River, Dene, First Nation) and a Métis community (Pinehouse), both in the heart of Cameco uranium mining operations. The third was in the northeast at a mining town (Creighton) that was predominantly Euro-Canadian but with another First Nations reserve nearby.

A remarkable grassroots movement emerged after Max Morin, Métis and a former member of the Royal Canadian Mounted Police, attended a First Nations Elder's healing circle when an official of NMWO was actually promoting the notion of a waste repository as a solution to adolescent Native suicide. The elders were extremely upset by this preposterous and very insulting appeal to tragedies that cause deep sorrow among Canada's Indigenous Peoples. In May 2011, Morin along with Debbie Mihalicz, a Dene couple Marius and Candyce Paul, and a group of Dene, Cree, Euro-Canadian, and Métis highly determined grassroots activists from the town of Pinehouse and English River First Nations organized a movement with the name—Committee for Future Generations—to immediately do everything possible to prevent such a repository and ultimately work against all things uranium and nuclear.

While there was a few Euro-Canadians involved, the movement took on a distinctly northern Native tone. For instance, Marius Paul often points out how a famous Dene seer warned her descendants to beware "the black rock" (i.e., uranium) as bringing destruction. Showing indomitable courage, this group of about 20 land and water keeper activists did not have an easy time in the politics of their home community politics. To some First Nations people and Métis, they were seen as a threat because the mining company Cameco provided one of the only sources of well-paying jobs at their mines. The official leaders in both communities were being offered various financial incentives for their regions in support of uranium activities and the possibility of waste storage.

That summer the Committee began an 800-kilometer walk from Pinehouse to the Provincial Capitol Building in Regina. They labeled the walk, "Walk for 7000 Generations," which was the approximate number of generations that would be at risk from nuclear radiation from any repository that was built in their territory. Along the way, members spoke in various communities and collected a total of 12,000 signatures against a nuclear waste repository. Water was collected at each community and mixed with that previously gathered at other places in a First Nations water ceremony. The meaning was to point out that water is the sacred basis of all life. Furthermore, all rivers, streams, groundwater, and aquifers are connected and would be in danger of radioactive contamination in any leakage since all water is connected and thus it would affect all of the potential water sources in Saskatchewan. After they reached Regina, several hundred marchers including some of us from a Coalition for a Clean Green Saskatchewan (CCGS) and other Native people in Regina joined them. At the Provincial Legislature, they repeated the water ceremony with a sample from Regina, held prayers by elders in both Dene and Cree, and then a series of speeches including several in the Indigenous languages and English. The ultimate object of the March and demonstration at the legislature was to present the petition to the Government and get a response.

For the next four years after the March for 7,000 Generations, the Committee for Future Generations collected another 10,000 signatures bringing the total to 22,000 and entered into public education campaigns in the North. We at CCGS had become their auxiliary associate and we shared our educational material with the Committee. Sixty percent of the residents of their home communities of Pinehouse and English River First Nation signed their petition after their diligent educational and lobbying activities, NWMO withdrew consideration of those areas in 2013 but to, perhaps, save face they claimed it was because they were not geologically suitable. Attention was then drawn to the Euro-Canadian mining town of Creighton in the opposite end of the province. There at the invitation of concerned citizens, both the Committee for Future Generations and CCGS gave joint presentations about the dangers of things uranium and nuclear. The Committee for Future Generations also presented their materials on the nearby Peter Ballantyne, Cree First Nation, and its Council later declared that any job opportunities that NWMO had touted were not worth the risks on land and water for future generations. In March 2015, NWMO announced that it had eliminated the possibility of Creighton as a site again citing geological reasons possibly to save face (see https://theecologist.org/2016/mar/30/sacred-land-unholy-uranium-canadas-mining-industry-conflict-first-nations for more information).

134 *Energy*

So, the success of the Committee for Future Generations was complete in preventing a high-level nuclear waste repository in Saskatchewan. The efforts of this remarkable and fundamentally Indigenous environmental movement have been momentous when you consider the power of those with whom they clashed. It was a battle against neo-colonialism and, perhaps at a minimum, inadvertent environmental racism with corporate and governmental interests seeking to place the world's most hazardous waste materials in First Nations territory where many still depend very much on land and water resources for survival.

Conclusions

Although historically favored and massively subsidized by American and Canadian as well as by other governments, the nuclear energy industry has been under siege since the category seven meltdowns at three reactors in Fukushima, Japan, in 2011. In the developed world, a number of former leading nuclear countries, such as Germany, Switzerland, Italy, and Belgium, are in the process of eliminating their reactors (https://www.cleanenergywire.org/factsheets/history-behind-germanys-nuclear-phase-out). The prices for other means of energy production such as renewables or natural gas continue to be ever more affordable than those for nuclear. For the few new reactors being built in the developed world, consistently huge cost overruns continue to be the norm, and the length of time required to build one seems to be significantly increasing—the average used to be about 8.2 years but increasingly in some cases to 30 years or more. The costs for reactors can range from $6 to $24 billion with some of the burden of delays based on stricter safety requirements since Fukushima. The overall portion of global energy provided by nuclear has slipped from 14% to 10% from 2010 to 2021 (https://ourworldindata.org/electricity-mix).

By now there are several hundred aging reactors reaching the stage where they must be decommissioned because it will be dangerous for them to otherwise continue. It is estimated by the U.K. Decommissioning Authority, that it would cost £131 billion to decommission Great Britain's remaining 19 nuclear reactors (https://www.gov.uk/government/publications/nuclear-provision-explaining-the-cost-of-cleaning-up-britains-nuclear-legacy/nuclear-provision-explaining-the-cost-of-cleaning-up-britains-nuclear-legacy). Added to the decommissioning of each country's nuclear fleet are the continuing costs of containing the nuclear disasters of Chernobyl and Fukushima and it would not be an unreasonable guess to assume the total global cost is considerably more than US$1 trillion. Then there is length of time required to complete the task for each reactor—50 years is suggested by the Canadian Nuclear Safety Commission (https://nuclearsafety.gc.ca/eng/resources/fact-sheets/decommissioning-of-nuclear-power-plants.cfm). A question remains regarding when will the sites achieve what could be considered "greenfield status"—being available for unrestricted activities could be allowed after all potential radioactive contamination has been removed. Presumably that would vary with the site. In the very north of mainland Scotland, the Dounreay nuclear facility has been undergoing decommissioning and the greenfield status is expected to be achieved in approximately 300 years (https://www.bbc.com/news/uk-scotland-highlands-islands-54085592)! Top all of this off by realizing that only one, Finland, out of the 30 countries making use of nuclear energy so far has an operational plan for long-term storage for its nuclear waste. All of this should make one wonder whether nuclear power has or should have a future considering all the costs and risks that remain at the end of the nuclear chain that have not been dealt with and may be the costliest of them all, let alone the current risks and costs all the way through it and those that have accumulated from past fiascos.

Another consideration of interest is that the industry is an example of what complexity scientists would call "path dependency" (see Ervin 2015)—once a particular path emerges or is chosen, it limits other directions that could be taken, grows in internal amplification of its characteristics, and generates certain unavoidable consequences. There are plenty of examples in human endeavor. As one, we can refer to our reliance on fossil fuels; the invention of the internal combustion engine and the car; their impacts on the design of human communities such as superhighways, suburbs, fast-food outlets, and shopping malls with resulting concrete-based construction sprawls that sacrifice valuable farmlands; along with the looming Damocles Sword facing us all through combined global climate chaotic changes and the possibility of extinction. Shifting from all of that has been the major challenge of the 21st century.

Marin Katusa (2012) writes indirectly hinting at another example of path dependency in his discussion regarding thorium as a potential nuclear fuel (although it should be noted that there is not a consensus on the viability of thorium). According to him, thorium is three times more abundant than uranium; with modification it can be used as a nuclear fuel; is less radioactive than uranium; it is not fissile on its own meaning that reactions can be stopped at any time; and its wastes are only radioactive for about 500 years. Yet when the first nuclear program began (and it is here we can surely all agree with him) its instigators were in a rush to invent an atomic bomb, which could not have been done with thorium. As has been emphasized in this chapter, the military and civilian dimensions of this industry have been essentially inseparable. Trillions of dollars have been invested, the vast majority of it public, and even when private investors become involved they are heavily subsidized by governments. The nuclear industry could not stand on its own, and as the earlier parts of this chapter have noted, its emergence came out of a desire to create the ultimate weapon that could kill thousands of people at once—not exactly a noble, humanitarian endeavor. Ever since, regardless of some positive outcomes such as the development of cancer treatment and medical diagnostic isotopes, the destructive potentials of the industry are implicit to its operations.

In spite of all of these negatives there are countries, notably China and India, that are building large-scale nuclear reactors in the context of their rapidly, super-charged, developing economies where all sorts of energy sources are in the mix including coal and renewables (https://www.world-nuclear.org/information-library/current-and-future-generation/plans-for-new-reactors-worldwide.aspx). Regarding those who are not just invested in keeping a dying industry alive because of their own professional commitments and/or strongly held and immutable pro-nuclear convictions, one could understand their motives in advocating or exploring the possibilities of using nuclear power as a carbon-free replacement for the sources of power, fossil fuels, that are leading the world to major climate crises. So what, indeed, are the possibilities of nuclear power being the salvation by ramping up to provide the energy needed to satisfy developing countries' needs as well as providing replacements for existing energy that are non-carbon emitting?

Australian electrical engineering researcher and professor Derek Abbot (2012) in an article in the *Bulletin of the Atomic Scientists* directly explores that monumental task. To begin with, he points out that the 2012 total of global electrical energy consumption was 15 terawatts. A terawatt is 1000 gigawatts, and a gigawatt is 1000 megawatts. Taking as an average a 1000 MW nuclear reactor (equivalent to the size of reactors proposed for Saskatchewan during the existence of the Uranium Development Partnership) or one gigawatt, he estimates that to replace the existing energy 15 terawatts that would globally require 15,000 nuclear reactors and at that time (2012) there were 430 reactors in existence. Considering the expansion required that is staggering. But there are many other factors including special conditions that they must be near large bodies of water for cooling purposes and not close to heavily populated centers in case of accidents. In the case of rivers, there is the danger of pushing temperatures far above acceptable levels with serious environmental consequences. In pointing out the United States in particular there were 60 nuclear sites (some containing multiple reactors), but the country would need 4,000 reactors to replace its required energy and finding even as few as a further 100 sites would be challenging. When the reactors are eventually decommissioned even after 60 years the territories upon which they exist would have to wait many more decades before achieving greenfield status. Each of these 15,000 reactors would have to be decommissioned after an average of 50 years of service. He calculates that in order to maintain the fleet of 15,000, one each of a reactor being constructed and one being decommissioned would have to happen each and every day. Considering the lengths of time and costs for each reactor that seems to be impossible. Then the hefty problem, as always, remains what to do with the waste and especially at this scale considering beyond the high level spent fuel *all* materials low and intermediate risk would also have to be buried. Accidents will likely proliferate on a major scale. He has calculated with 14,000 years of active output spread over the lifetime of 580 reactors that there were 11 major accidents of a full or partial core meltdown. So if there were only one reactor in the world that would mean one serious accident every 1300 years. With 15,000 reactors that probability would average out to a serious accident every month. Then there is the issue of available uranium. The estimate at the time was that there was enough available for the existing reactors for another 80 years—if scaled up to 15,000 reactors, there would be less than five years of supply available. He also deals with fast breeder reactors, the heightened dangers of nuclear risk regarding nuclear weapons and terrorism, and the massive loss of rare and valuable materials that should be recycled but would be lost for use in the burials of nuclear waste.

So obviously the possibility of a nuclear renaissance solving our most pressing environmental problem, climate change, especially in the very short time required is a utopian impossibility. Abbott (2012) instead recommends renewable sources of solar and wind supplemented by natural gas.

Then, finally, consider the situations of the people in the context of front lines of weapons testing, lands being used in mining, or as actual or potential sites for nuclear waste disposal. In this chapter we have considered the circumstances of cases from two indigenous peoples. The case from Deline, NWT, clearly shows very negative environmental and social impact consequences (http://www.ccnro.rg/dene.html). Regarding the *Dënesųłıné* (formerly referred to as Chipewyan) of Northern Saskatchewan there has been no extensive and thorough investigation of impacts and the provincial government, in spite of repeated requests, has consistently avoided doing a baseline health study in the region from which any increased prevalence of cancer and other radiation illnesses could be measured. Considering the rest of the world where uranium mining is being done, the vast majority of it is done in areas that are the homelands of native peoples in the hinterlands of nation-states where such peoples tend to be the victims of colonialization and marginalization (see especially Johnston ed. 2007).

I think it is safe to say that, in general, this industry in all its manifestations has practiced variously what can be termed as wastelanding or environmental racism in what are ostensibly sacrifice zones. And then ponder the implications of this quote, "If Julius Caesar had established a nuclear reactor in Rome, we would still be looking after its waste" (Anonymous).

Part III
Solutions

10 Transforming our Political Economies and Dealing with the Issue of Constant Growth

This chapter has to be considered pivotal to any solutions to climate change. It would be unavoidable to conclude otherwise *that they have to address the current dominating political economy—capitalism especially in its neoliberal form--and the assumption of the need for continuous growth.* That is not to let soviet-styled communist regimes off the hook--because they too were "productivist" or maximally growth-oriented, but such entities have disappeared. are marginal, or have combined capitalism with their state planning. Most causes for all the issues discussed in this book and beyond can be traced to the massive extractions of materials wrenched or aggressively coaxed (as in the case of contemporary agriculture) from biophysical environments--our essential life support systems--then processed and manufactured into commodities that are consumed in immense quantities but overwhelmingly in developed countries and through burning colossal amounts of carbon-dioxide emitting fossil fuels.

This is all framed by capitalism's fundamental orientations to accumulation, its profit motives, *and, again, its never-satisfied need for continuous growth.* The *problems, then, are most obviously directly caused by the operations of our political economy.* So, in response there have been three broad styles of proposals that attempt to deal with the economic side of the dilemmas.

Green Growth advocates, while making proposals for some changes and mitigation, still promote the power of existing market and financial investment principles, economic infrastructures, and institutions to shift directions to "net-zero" carbon emissions while allowing the global economy to continue its growth. Those supporting the notion of a *Green New Deal* continue somewhat with those ideas but advocate much more concentrated government regulation, intervention, and investment—all by analogy similar to Franklin Roosevelt's New Deal of the 1930s. The third proposal is very much more radical—it advocates major *Degrowth* or shrinking of the economy especially as focused on the already highly developed countries of the Global North while encouraging moderate but measured economic growth in the Global South so that, in the name of social justice, people there can come to experience sufficient degrees of comfort, health, and security that have been taken for granted in the North. There is continuity with current economic principles with the first two but a major breach with the third.

While I will outline these three approaches, it should be noted that space does not allow me to fully expand upon the many issues of debate among them. Some of these issues include is the particular approach too inflationary; will it cause major recessions; what about the need for jobs; how and who will pay for it; does it involve governmental, top-down, command economies; or an over-reliance and trust in the private sector that got us to the crisis in the first place; how will the debt generated by solutions be managed; can we trust the private sector to do the right thing and avoid the temptation of short-term profit and over-privileging investors; and so forth? The reader could explore these issues on her own by delving deeper into the literature.

Green Growth

With some adjustments the notion of Green Growth as the solution makes use of the existing global and national economic systems and has powerful official backing.Daniel Fiorino (2018) characterizes Green Growth as a pragmatic agenda to manage climate change and other environmental issues. Prosperous societies, suggests Fiorino (2018), after having gained from economic growth, are prone to sharp declines in population growth, and with improved social conditions such as in the status of women, are likely to "decouple" their economies from dirty, heavy industrial and polluting industries and engage much more in environmental protection. His view is that ecological stewardship and economic success are most certainly not incompatible.

Green growth advocacy has become more noticeable as a recommended policy direction as a result of the 2008-2009 Recession, because capitalists needed to restore faith in the financial markets and encourage renewed economic growth at the same time. It dominates discussions at the World Bank, the OECD, and the United Nations Environment Program. These institutions and others have underscored the still pressing need to globally raise many more people out of poverty. The necessity for economic growth is still seen as essential by proponents and now urgently coupled with the even more pressing requirement to rapidly reduce the risks of irreversible climate change. They also point to the inescapable fact that population is going to continue to grow, thus the provision of goods, services, and development will have to grow as well. Capitalist mechanisms such as the market and private investment, they claim, have proven to be the most effective ways of accomplishing all of this.

Green growth approaches, however, do take into account the needs for technological innovations leading to greatly reduced, even ultimately "net zero" carbon emission, and the need to recognize and incorporate as the costs--externalities or damages caused by polluting and carbon reliant industries. For that reason, proposals rely on some degree of government intervention through interventions such as pollution taxes and carbon cap and trade systems. In the latter, governments put graduated, increased limits on the total CO_2 emissions allowed and then let companies financially trade amongst themselves for the rights to temporarily exceed the limits determined for them but only within the total allowed as that shrinks over time to nothing or virtually nothing. The rewards thus go to the industries that decarbonize the fastest.

Financial mechanisms need to increasingly reorient the market towards investment in green industries. The framings of the policy and business agendas should favor economy-ecology relationships where there had previously been lacking. Renewable energy is the obvious example and one that increases the likelihood of both environmental and human health, and the industry is expanding very rapidly, more so than other sectors of energy, so further investment already ensures good returns from investment.

Although he does not specifically label himself as such Mark Carney (2021) is clearly an advocate of Green Growth and working within the existing system where capitalism thrives.

> My experience has made me a profound believer in the market's ability to solve problems, I have seen day in and day out the very human desire to grow and progress, of people's yearning to make better lives for themselves and their families. Continued growth isn't a fairy tale; it's a necessity (Carney 2021: 338).

Carney is highly influential having served as the Governor of the Bank of Canada where he navigated the country out of the worst impacts of the 2008-09 Bank Crisis and Recession; then he became the Governor of the Bank of England similarly leading the United Kingdom through the financial instability that could have ravaged the country after Brexit.

His book *Value(s): Building a Better World for All*, is extremely wide ranging, and, as an economist, he feels the necessary task is to link economies with their societies' values. In general, he claims, markets and prices to determine value have done a relatively decent job of linking to such values (e.g., generating prosperity and provisioning people's material needs) but they did much more so in the past than now. Recently there have been disconnects that have lost the confidence of the public so that there needs to be a nudging of the market and financial institutions towards issues that are now increasingly valued by society. The public desire to solve the climate crisis is one of those shifts, along with growing senses of the need for social solidarity and more equality in the distribution of wealth, and the striking necessity for much more resilience in major risk areas such as pandemics and climate-generated disasters. These are all values that he feels can be accommodated by capitalism without the loss of economic growth.

In fact, he sees them as enormous investment opportunities and of high significance because the required transformation to adjusting to climate change is "job-heavy" at a time when under-unemployment has also become a chronic problem. One of the things that he feels is a major asset of capitalism is its historic capacity to generate and muster innovations to problems needing to be solved. That is essential for the major task of decarbonizing electrical production, transportation, and manufacturing. The focus has to be on achieving *net-zero carbon emissions* through mandatory but manageable quotas and stages. The implication of net-zero is that some carbon sources such as biofuels or even some uses of petroleum and coal could be balanced by the use of carbon-capture technologies or by the fact that growing replacements crops for any biofuels burned will be recaptured through their natural growing and decaying processes.

Without the private sector playing *the* major role, he insists, the climate change crisis will not be solved. This is because the investment required to bring down carbon emissions to net zero will be about $3.5 trillion a year for many decades just in the energy sector alone, and because only the private sector has that kind of money. Furthermore, he maintains that the private sector is superior when it comes to innovation, initiative, and management.

The public sector does have a major role in steering the transformations. Strict and meaningful carbon taxes and cap and trade systems will need to be legislated. Governments, themselves, will have to invest in some major infrastructure projects. They should provide grants, loans, and tax reductions to support the research and development towards sustainable technical solutions. A broad consensus among the leaders of nation states has developed that global temperature must be kept under 1.5 degrees Centigrade in temperature rise and that rapid change in economies through technologies must be achieved. For that there is a need to "…. orient (impact) investing toward that goal…" and that "…investors will focus on how well companies are positioned to manage the associated risks and opportunities (Carney 2021: 449)." Through disclosure devices that he recommends, companies will reveal the amount of assets that they have devoted to net-zero carbon projects among other relevant revelations. In keeping with his values approach, he envisions a "mission-oriented" capitalism.

Governments will have to generate regulatory policies that frame future directions—such as precise-scheduling in the phasing out of internal combustion engines, among other reasons, to provide certainty to the private and investment sectors as to their commitments. Governments can assist in the leverage of all this by promoting activities that are both lucratively job heavy and capital intensive and help to place their countries in competitive positions internationally. Policies should be made tangible for citizens and businesses so that they are encouraged to make their homes and buildings energy efficient and smoothly make the shifts to driving electrical vehicles, Regions that are hard hit by the transformations (presumably jurisdictions such as Texas and Alberta with its tar-sands) should be given special assistance in making the difficult transition to a net-zero carbon economy. There will still be some place, though, for petroleum production with low carbon technologies and carbon storage facilities.

He supports a rapid transformation from internal combustion vehicles to electric ones along with the establishment of a vast network of charging systems, accompanied by subsidies to compliant manufacturers and purchasers. Investment, grants. and subsidies also need to be directed to the development of hydrogen fuels and for carbon-capture technologies. Making housing and other building structures green needs heavy investment and would create many new jobs not only in their refurbishment but in ongoing construction of new stock. Again, as the main driver for mustering the changes Carney relies heavily on high and constantly rising carbon taxes that are clearly defined and scheduled so that private firms are aware of the changes they have to make and are otherwise" incentivized" to participate in this massive transformation that should coincide with the 4th Industrial Revolution that will involve much in the way of artificial intelligence, robotics, and nanotechnologies.

Green New Deals

The essence of Green New Deal (GND) formulations is to bring about the needed change by bringing massive amounts of government resources to work while providing direction, incentives, and partnerships to the private sector. *In contrast to Green Growth, the principal driver would be government not as much the market.* There would be frequent legislation that mobilize money, time, people and institutions to bring about as complete a transformation of each country's energy systems to non-carbon production in the shortest times possible—ultimately bringing about a transformation that creates a new economy. It means the major overhauling all of nations' infrastructures involving upgrading and expanding integrated, electrical grid systems; generating renewable power to replace carbon sources; establishing major forms of public transportation such as high-speed intercity railways run on renewable energy systems; and directing and coordinating manufacturing interests in the private sector to produce the materials needed to accomplish all of this by setting national goals while generating the necessary incentives and penalties.

There would be precise goals, laws, quotas, and target figures to reduce pollution of all kinds and most especially carbon dioxide emissions. Government funding of research devoted to green technologies, land and habitation restoration, and other green solutions would be extensive. Similarly, substantial efforts through public funding and incentives to private firms for job training and retraining for the new green economy would be revolutionary. Young, student, and unemployed labour would be mobilized into large civilian corps to engage in such things as conservation, habitat restoration, and assisting local communities in activities meant to build resilience in dealing with the consequences of climate change.

The idea is to emphasize the notion that nations, their citizens, and the entire global community face threats *right now* that are of enormous magnitude, and that we should be on a crisis footing similar to American New Deal responses to the Great Depression of the 1930s. The emergency metaphor is "all hands on deck" in the sense that a GND requires everybody's support and participation. To reinforce that, certain aspects of social policy, especially if they are already lacking in such nations, should be added to the legislation packages. These could include universal healthcare, free or low-cost university education, and basic, guaranteed incomes above the poverty lines. Any underprivileged, neglected populations with historical grievances such as Afro-Americans, Native Americans, and Canadian First Nations peoples should be given remedial advantages in participating in the GNDs especially since they have so often been the victims of environmental racism.

Seth Klein (2020), a Canadian proponent for a GND titles his book *A Good War: Mobilizing Canada for the Climate Emergency*, makes use of Canada's experience in World War II as the example of historical precedence for such concentrated, national mobilization. At the time of the War's outbreak, Canada was an immature nation, over-dependent on its natural resource industries, with a tiny insignificant military, with a small industrial infrastructure, and still tied in many ways to the foreign policy dictates of the "mother country" the United Kingdom. Rapidly the federal government organized all the federal departments towards a united war effort. Over a million men and women out of a population of only 12,000,000 enlisted and were supplied for military service. Canada's navy and air force were amongst the largest in the world, and Canada's army played a major role in liberating Western Europe. The nation's industrial structure was massively expanded with the cooperation of the private sector, and many thousands of tanks, trucks, small arms, artillery pieces, fighter and bomber planes, warships, and merchant vessels were produced not only for its armed forces, but for Britain, besieged by NAZI Germany, was supplied by Canada with manufactured goods. Canada's agriculture and nutrition standards were organized federally with food rationing and with significant shares being directed to feeding the military well and exports to help provision beleaguered Britain which had been cut off from its usual sources by German submarines. The civilian population was mobilized and in solidarity with the war efforts in spite of the sacrifices demanded of it. Seth Klein's (2020) purpose in the book is not meant to raise jingoistic pride in Canada's history but to show through historical analogy of how "we can do it" promoting a Green New Deal and that as with World War II likely come out the experience better for it. The underscored point being that. the threat was so large and unthinkable, that of the NAZIS winning the war, that there was no choice but to unite and strive for victory. *The threat of climate change, in the opinion of Green New Dealers, is even greater so the time to mobilize is now.* Metaphorically we need to put ourselves on the equivalent of a war footing overcoming a crisis that must be solved.

Globally, as Seth Klein's sister Naomi Klein (2019), one of the GND's most passionate advocates, suggests GNDs can, on the positive side, create hundreds

Political Economies and Dealing with the Issue of Constant Growth   147

of millions of new good jobs. Focusing on social justice, among other things, to reinforce commitment and solidarity in dealing with such a major crisis, many of the jobs and enterprises required would go to marginalized peoples and communities. Overall GNDs would also support extensive investment in health and education services with an emphasis on maximizing the "planet's life support systems" (Klein 2019: 26) in more ways than restoring some of the balance in ecosystem viability. With its concerted public efforts, it would go a long way in fixing an economic system that has failed people in so many ways.

She sees some variations off of the tone and character of the original American New Deal. While changing the energy productions systems away from fossil fuels as much as possible, the new renewable energy projects should be locally, community owned and managed rather than centralized, monopoly-oriented, and top-down as with the massive coal, hydro, and nuclear plants that have powered the current era.

Jeremy Rifkin (2019), who has acted as an advisor to the European Union on GND issues, offers twenty-three specific policies for implementing them. I have paraphrased and reduced the content and generalized them beyond the specific U.S.A. context that he has devised them for.

1. Across the board government imposition of carbon taxes with revenues to support other GND policies and rebates for the vulnerable so that they receive more than they would have to pay.
2. A rapid end to all government subsidies to petroleum industries.
3. Creation of a national integrated smart power grid that accommodates a rapidly emerging set of green, renewable energy sources, all for electricity.
4. Establishment of tax credit systems that help to rapidly support innovations and the development of green energy electrical technologies for that grid and that lead the nation toward zero carbon emissions.
5. Tax credits and other incentives to be given for the development of energy storage systems including those at commercial and household levels for managing the viability of the integrated smart power grid.
6. Broadband Internet access to be distributed to rural and marginalized communities for environmental and health improvement information generally found with wireless connections.
7. Industries using data centers should receive tax credits for installing renewable energy source to power them.
8. Major tax credits to be given for the purchase of electric vehicles while graduated and increasing tax hikes are established for ones using internal combustion. 2030 should be a cut-off date for the curtailments of all cars, trucks, construction machines, and buses using petroleum.
9. To assist in this, tax rebates should be provided for the establishment of electrical recharging stations in and around residential, industrial, and commercial sites.

10. Governments should mandate that all their properties be established to produce zero-emissions of carbon and all its be procurements be green in quality to support green business.
11. Governments should encourage and develop means for phasing out petrochemical-based agriculture--promote organic means and reinforce regional and local markets with 100% organic certification by 2040.
12. Government inducements for farmers to use carbon storing farming techniques and to transform marginal land into carbon capturing wildlands. As well governments should reforest public land for carbon-capture.
13. Water and sewage systems as well drainage infrastructures should all be upgraded to be resilient to climate change.
14. The mandating of circular, presumably recycling, dimensions into all industries and supply chains minimizing waste materials and pollution.
15. Train and make use of the military as first responders to climate change crises and to participate in restoration projects.
16. The establishment of national green banks from which local banks can draw to fund state, provincial, and other mote local entities to borrow from for infrastructure related switches to renewable energy systems.
17. Such green banks could be primarily provisioned with investment capital from both private and public pension funds with approval from the unions involved. Priority should be given to union workers in such project and the rights of yet to be organized workers in such projects are legally assured.
18. Younger people at the student level need to be assisted with government funding to learn the new skills required in the rapid climate change regime. Using American examples, he also suggests the creation of such entities as Green Corps, Conservation Corps, Climate Corps on the analogy of the previous Peace Corps, and Civilian Conservation Corps of previous era to employ high school and university graduates in public service to asset communities impacted by climatic change and other environmental challenges.
19. Governments should prioritize green new deal business opportunities in disadvantaged communities providing locals with the kinds of training to build on such opportunities.
20. Tax reforms need to ensure a more just and equitable society reducing vast disparities between the super-rich and other sectors of the society and to assist in the transition to green economies.
21. Governments need to prioritize funding that reinforces Green New Deal policies, technologies, and transitions.
22. Time frames need to be established, quickly set in motion, and succeed in the above proposals such as renewable energy development, extended and seamless integration of national grids and Broadband communication links etc. Along with all the codes and regulations needed to maintain these transformations.

23. International cooperation needed to extend the benefits of a Green New Deal with special emphases on cooperation among the U.S.A., the European Union, and China.

An interesting dimension of Green New Deal proposals is that both Rifkin (2019) and Seth Klein (2020) recommend citizens' councils and forums at regional, local, and national levels for ordinary people to discuss, challenge, revise, and present alternatives during GND campaigns. It is truly important for them to have bottom-up capacities to cultivate the importance of participatory democracy and solidarity in dealing with global, climate change threats.

Degrowth

Kallis et al (2015:3) describe the degrowth alternatives in clearly radical ways as a collective "decolonizing of the public debate from the idiom of economism and the abolishment of economic growth as a social objective." According to them, there is urgency in that the direction has to change to one where societies use very much less quantities of their natural resources; find new ways of managing them without so much waste; devise more inclusive ways of sharing technology and commodities; and design facilities with far less emphasis on private property. People will have to live frugally most especially in nations that are considered highly developed. "Simplicity", "localism", "conviviality", "caring", and "sharing the commons" will become important keywords and phrases in the design of this new form of society. Major downscaling of the production and consumption processes are essential to degrowth—massively reducing energy consumption and the end of *all* fossil fuels. According to Hickel (2020b), it means a major reduction in material throughput and bringing *an economy to a balance with nature* while at the same time improving human well-being and reducing inequalities. In his critique of capitalism and the present growth economy, he points out that the growth touted in annual reports of increased Gross National Products is false propaganda, because the gains virtually all go to the elites in society.

A major transformation is required—the metaphor used by Kallis et al (2015) is *not creating a leaner elephant but instead a snail*. With this image of a social entity with a metabolism, it would require much less energy and have intrinsic ways of recycling much in the use of any materials and make multiple uses of that same energy (e.g., power and heat simultaneously). Activities will be radically changed with different gender arrangements, allocations of time, paid and non-paid work, and, most crucially, different relations with the non-human world. Work sharing, cooperatives. and both basic and maximum incomes guaranteeing everybody a livelihood and a decent but far from excessive standard of living. Of the three broad orientations Green Growth, Green New Deal and Degrowth, Degrowth formulations make the maximum case for radical transformations while boosting equality and social justice.

150 Solutions

The supporters of Degrowth find the term "development" problematic for economic and social planning. Its basic meaning is found in biology where an embryo ultimately becomes a mature adult, but the growth or development for the organism then stops, and it operates to maintain a steady-state relationship with its environment. So too should societies and their economies (Kallis et al 2015). That indeed was roughly true of so many societies, regionally and traditionally rooted, as anthropologists have demonstrated quite thoroughly (see Sahlins1972). This was much more of the case previous to the Age of Discovery, the spread of capitalism, and the universal imposition of the power by the nation state over previously autonomous, indigenous peoples.

Degrowth economic ideologies harshly condemn the market and capitalist systems that depend so much on money as a measurement of value and as the medium with the power to generate action. The critique of capitalist modernity laments the reliance on the Gross National Product and its constant growth to indicate progress, and the commodification into money terms of just about anything considered of value. The market has come to dominate so very much of the power that allocates resources and so much of that is misallocated and creates damage. Consider these observations by Joshua Farley one of the founders of Ecological Economics:

> Markets allocate resources among products and resources in a way that maximizes monetary value. The question, however, is whether monetary value is actually what we want to maximize. If an American is willing to pay more for corn to make ethanol for her over-sized sports utility vehicle (SUV) than a malnourished Mexican can afford to pay for tortillas, then converting corn to ethanol maximizes monetary value. Markets allocate resources based on the principle of one dollar, one vote and *future generations have no vote. Markets are guided by the preferences of living individuals weighted by their purchasing power.*
>
> Furthermore, markets rely on negative-feedback loops. As resource scarcity increases, prices rise, signaling consumers to consume less and suppliers to supply more or develop substitutes. Prices balance supply with demand. However, the supply of low-entropy matter-energy and land is fixed. When the price of non-renewable resources like fossil fuels or mineral increases, we may extract in-ground stocks more rapidly to temporarily increase current supply, but at the expense of future supply. If we extract renewable resources like timber or fish more rapidly in response to a price increase, we may actually decrease their capacity to reproduce, again reducing future supply. It is also extremely difficult to develop substitutes for fossil fuels, land, and ecosystem services.
>
> (Farley 2010:269)

Farley's senior colleague, Herman Daly (Daly and Farley 2011) in another context (Daly 2015) suggests a number of policies that would help to support an economy much more in alignment with Nature with its limited stock of resources to support humanity, other beings, and future generations. Relative steady state, rather than perpetual growth, economies would be the ultimate goal of Degrowth advocates after making amends for glaring inequalities.

One of his most intriguing ideas is *ending the fractional reserve banking system*. That system allows banks to lend out vast sums, in some cases with the actual assets on hand in support of the money lent to be as low as only 3%, thus creating ever-increasing amounts of money as credit. This is done digitally and not through actual currency, thus as it were "out of thin air", and lent at the rate of 97% beyond the actual holdings on reserve in the bank. This constantly increases the glut of debt that inflicts the world today. In return, real actors in real time then must repay the debts by actual economic activity, extraction, material throughput, processing, manufacturing, and consumption—all keeping economies growing at ultimately exponential rates. Banks, in his view, should be restricted to lending at 100% reserves thus limited by the assets that depositors have actually made not the fictive excess that so very much disturbs the economy and Nature, and generates so much inequality.

As another proposal, personal incomes would be based legally on guaranteed, minimum standards of well-being and also limited as at the maximum amounts to create just enough incentive for innovation and enterprise but low enough to prevent a plutocracy of the super-rich. Taxes would shift more from the "goods" to the "bads" to greatly reduce pollution, CO_2 releases, and wasteful inefficiencies in material throughput. Limits and caps would be placed on renewable resource extraction so that they would not be depleted beyond their capacities to regenerate. Non-renewable resources should not be depleted faster than renewable resources could be developed to replace them. Wastes from both renewable and non-renewable resources should not be returned faster than they can be reabsorbed by Nature. Global governance systems need to be developed that curb the excesses of corporations more recently freed from the regulations of nation states in their misuses of resources.

In keeping with both anthropology and the political ecology that constitute the core of this book, it should be noted that degrowth ideas have been supported by several prominent anthropologists. Significant is the endorsement of Arturo Escobar who, as I mentioned in Chapter One, is one of the key founders of anthropological political ecology as well as a major critic of development (Escobar 1995). He has become involved with Postdevelopment movements most notably in Latin America where they are thriving especially in connection to the non-capitalist, nongrowth, Indigenous peoples', communalistic aspirations and their local ontologies (overall perspectives on existence and being). These ideologies vividly contrast with Western ones of capitalism, dualism, extractivism, individualism, and materialism that have external imposed themselves and created so much damage in their lives since the Age of Discovery and colonialism.

In a paper titled "Degrowth, Postdevelopment, and Transitions: A Preliminary Conversation" (Escobar 2015), he discusses the convergences with Degrowth thinking that has been emerging in the developed world with the Postdevelopment movements with which he has been engaged. In keeping with his own post-structuralist tendencies, he views them as *"transition discourses"* that are preparing us for the necessary transformation away from the hegemonic damage of Western capitalistic styles of development and massive extractivsim that are the sources of the overall crisis. There is a striving for radical cultural and institutional transformations—transition from "oil to soil" –transitions to "decentralized, biodiversity-based, organic food and energy systems operating on the basis of grassroots democracy, local economies, and the preservation of soils and ecological integrity (Escobar 2015: 453.)."Away from growth so enthusiastically touted in the capitalistic development model, he briefly, through Buen Vivir and the Rights of Nature movements, discusses similar "de-economized " views of life inspired by Indigenous concepts and ontologies in Latin America as well as the agroecological approach in producing food. Collectively the many Degrowth and Post Development strands of thinking "…agree on the fact that markets and policy reforms, by themselves, will not accomplish the transitions needed (Escobar 2015: 457)". One significant takeaway is that instead of the hegemony of the capitalistic, development model, there should be a plurality of alternatives— a *"Pluriverse"* as he puts it. Throughout them, social solidarity, communalism, community, sharing, and relationship with consumption patterns shifting to quality of life rather than accumulation, are underscored. Escobar elaborates on these proposals in Escobar (2017) and Kothari et al. (2019).

Degrowth, advocate Jason Hickel (2020a), a prominent British economic anthropologist with extensive knowledge about global development, provides a harsh evaluation of capitalism and its obsession with growth. Among other things, he demonstrates that there is no linear relationship between increasing growth in personal economic terms and happiness and well-being. Studies in developed countries have repeatedly shown that there is a saturation point after which additional income or wealth makes no difference.

Approaching the history of capitalism from an anthropological perspective, he suggests that it created a brand-new ontology or world view that generated false senses of scarcity that could only be met through accumulation, private property, employer-imposed disciplined wage labour, and the constant growth of wealth through markets that were obsessed with exchange value rather than on use value. Nature was commodified and relationships among the parts become purely utilitarian leading to the kinds of damage we have seen over the last several hundred years. Previously, world views tended to emphasize interbeing, personal, sometimes even sacred relations among the parts, other beings, and humans. People in non-state societies could comfortably subsist according to their regional ecologies without wage labour and ever-expanding economies.

Political Economies and Dealing with the Issue of Constant Growth   153

He also succinctly and tellingly demonstrates the huge discrepancies in the use of materials in their throughput as well as energy use. The Global South pays the highest prices in terms of existing environmental damage and climate risks, and the Earth as whole has been suffering major overshoots in material uses. Scientists have estimated that materials extracted have a limit of 50 billion tons per year to maintain sustainability, but as Hickel points out that the figure is now 100 billion tons. High income countries such as the USA and Canada are consuming about twenty-five tons per capita, while low-income countries only about two, and middle-income countries twelve. To meet sustainability targets, the figure should consistently be about eight globally. Then considering per capital emissions in carbon dioxide, similar patterns emerge—1.9 tons for India, 8 tons for China and 16 tons for the USA.

What is even more staggering is the historical basis for the accumulation of CO2 in the atmosphere whereby the Global North that represents only 19% of the total population has contributed 92% of the accumulated overshoot with the USA accounting for 40% of the total. In contrast the entire continents of Latin America, Africa, and the Middle East account for 8%. Tellingly, Hickel refers to all of this inequality as a form of *"atmospheric colonialism"* – "A small number of high-income nations have appropriated the vast majority of the safe atmospheric common, and have contributed the vast majority of emission in excess of the planetary boundary (Hickel 2020a:115)". Then with regard to the ongoing costs of climate breakdown, the South in 2010 bore 82% of the $571 billion cost through droughts, floods, wildfires, landslides and storms and from the same study he points out that 98% of the deaths occurred in the South.

Hickel advocates shifting more of the world's wealth and income to the Global South, and among the mechanisms to do that would be to cancel *all* the debts imposed by institutions such as the IMF and the World Bank. That would allow these developing countries, instead of having to deal with burdensome annual debt payments to spend more on health and education and other services benefitting their peoples. He would also put an end to *all* the foreign land grabs and redistribute them back to the people, usually poor, from which they were taken in the first place. Removing the massive subsidies that enrich farmers in the USA and the European Union would allow farmers in the Global South to able to fairly compete in international markets.

Not only does he point out that individuals with high incomes after a certain saturation point are no happier and content with their lives than those of a modest income, this applies broadly to countries as well and has lessons for "development" without significant growth. In contrast to the United States with the highest GNP and per capita GNP, he points to the successes of Finland, Denmark, Cuba, Costa Rico, Kerala in India, and Sri Lanka with less or much less in total and per capita that have, nonetheless, provided healthy meaningful, equitable lives for citizens with emphases on social policies.

Hickel (2020a) offers these broad proposals to further help fashion a degrowth economy:

1. **Legislate against planned obsolescence**—For the last hundred years manufacturers have fostered built-in lifetime limits to items such as light bulbs, automobiles, refrigerators, and just about everything else. This encourages steady rates of consumption favoring the manufacturers and their profits—but adding substantially to material throughput, energy use, pollution, and waste. Their failures are usually due to the strategic placement of small electrical components that could, otherwise, be replaced to last many times longer. Many products are also designed to prevent mechanics or the users themselves from repairing them so when the product reaches its expiry date, it is much easier to simply throw them away and buy new ones. Supplementary products such as connecting wires, ports, and external disc drives are also turned over rapidly in design such as in computers so that older parts cannot be used in newer products or vice versa. These purposely standards of inefficiency aiding in the establishment of business profits mean massive throughputs of materials and energy uses that most certainly extend our climate and environmental crises and continue to generate and exploit the extension of the consumer culture—making its users even more helpless. Among the solutions that Hickel suggests are mandatory warranty guarantees that last ten years or more and "right to repair" laws that require manufacturers to make all their products accessible to repairs.
2. **Drastically cut the range and content of advertising**—Advertising is yet another mechanism to cause the consumption part of the economy to continue to grow by psychologically playing on people's insecurities. Once people bought what they needed--they now tend to buy what they have come to believe what they want and those wants in many cases are artificially created. What could be done? Public spaces for advertising should be banned. The total amounts allowed for advertising could be limited with the establishment of rationed quotas. Laws could be directly aimed at eliminating the most manipulative types of advertising.
3. **Shift from duplication in ownership to usership**—There is a lot of inefficiency and unnecessary duplication with capitalism's emphasis on private property especially with regard to personal and household items such as lawnmowers and power tools being used rarely and then only for brief times. He points to the existence already in some communities of neighborhood workshops where people can borrow or rent such equipment. These cooperatives could also be scaled up and supported by larger municipalities. This could reduce significantly the throughput of raw materials in manufacturing. Probably the most extensive example with regard to this issue is the private ownership of automobiles. The way around this would be to develop massively organized systems of public transportation and the

making of extensive networks of safe bicycle lanes in cities. Ride-sharing systems needed to be extended in operations that are not as exploitative of the worker/owners as the current Uber-like systems.

4. **End Food Waste**—As much as 50% of the food produced annually, Hickel claims, is wasted—2 billion tonnes. Conservative figures of at least 30% could at least hold true as this book's chapters on agriculture confirmed. The issue in developed countries is the high expectations of cosmetically perfect displayed produce, and unnecessary best-before dates, leading to massive amounts of food being thrown away. This is exacerbated by the long distances required in supply chains bringing food from farm gate to table. Then food is further wasted rather than recycled when it is allowed to rot in landfills. Hickel sees food management as a "low-hanging fruit" with regard to degrowth and remedial action on climate change. Laws could prevent food waste through mandatory requirements to give unsold produce to food banks. All food waste could be mandated to composting and recycling functions. Laws that direct and mandate localization making sure that the most significant bulk of at least staples are produced within the regions where they are produced would help a lot.

5. **The scaling down of ecologically destructive industries.** Raising beef is the first example that Hickel gives. The need to restrict that industry considering its demands for land, water, and such a high proportion of the globe's production of grains is mandatory if we are serious about our misuse of agricultural lands and global warming impacts of methane releases. Shifts to plant proteins such as beans and pulses would take pressures off and allow much land to be restored to forests and other types of wildlife preserves serving as additional carbon sinks. Scaling down the arms industry and the military would benefit everybody except arms merchants and those clinging to power while freeing resources for vital climate and environmental remediation. Ending the excessive use of plastics—especially as related to single use items—would be a significant contribution. The air travel industry needs to be restricted in its growth and ordinary operations since it has the highest CO_2 impacts per mile travelled. Scaling down on the non-essential products that we consume would lead to the reduction in the materials and costs of factories that produce consumer products. This leads to much less need for dealing with waste products, pollution, and the energy required to produce all this activity and material throughput.

A significant argument and agenda for Degrowth is provided by French engineer Philippe Bihouix (2014) in *The Age of Low Tech: Towards a Technologically Sustainable Civilization*. He points to the very dangerous annual overshoots of the Earth's resources and reveals that eighty percent of the total consumption is done by the twenty percent already most prosperous members of humanity. Alarmingly, in just one--the recent--generation, the equivalent of all the minerals that were ever mined were "scraped from the earth's crust". Continuous and even further expanded mining would be needed to accomplish the transformations and subsequent continuation of any green new deal or green growth plans. Such extractions and manufacturing to create a global renewable energy regime are likely to accelerate the very climate crisis that they are meant to prevent! Plus, the materials required can never be fully recycled so the problem will continue indefinitely. There are many other material situations that have been pushed to the limits by the high technology/industrial/high energy systems now in place. Bihioux advocates and documents pathways to societies based on simpler, less powerful technologies but having resource efficiencies and that are controlled locally. A complete rethinking of human needs is required with an orientation towards "sufficiency" or "just enough".

The Degrowth approach to dealing with climate change, multiple environmental issues of injustice, loss of biodiversity, degradation, pollution, the loss of life-enhancing support systems, equity, and the preservation of what is needed for future generations, has much to offer. Yet the chances of the proposals being implemented are very slim to not at all when we consider as Vaclav Smil (2019) points out that there are no political entities in the world or any serious political parties, especially in developed countries, that support Degrowth. It appears then, insofar as Degrowth might be practiced, at least in the short term, it will be experimental within some regional social movements and intentional communities.

The obvious impediment is the cultural values of modernity—it would be hard for anybody to countenance an actual decrease in their standard of living. For Degrowth to catch on as a meaningful driver of policy there would have to be some extremely major changes in belief systems and world views and that would be widespread enough to transform whole cultures and so many of the assumptions of modernity. Or, as I suspect, there may have to be drastic changes in the living circumstances of people to enforce such transformations—that will be explored in Chapter 14.

Politically Who is Going to Take Responsibility for Solutions?

What about the political side of the concerns for solving climate issues? Who has the authority and power to follow through with policies that bring about effective results?

The seventeenth century philosopher Thomas Hobbes devised the peculiar metaphor of *The Leviathan* (1651) or the all-powerful sea beast featured in the biblical *Book of Job* to represent the sovereign right of kings over their subjects. It was his view that such an arrangement of power was necessary to protect the lives and well-being of their subject who in turn provided obedience to the sovereign ruler to their projects. This pioneering work of political philosophy was central to the emergence of the nation state as a consolidated political power with

Political Economies and Dealing with the Issue of Constant Growth 157

its monopoly on violence. This was an implicit social contract—in exchange for their obedience, the subjects (later citizens) would not be endangered by random violence and disorder but instead experience peace and security. Hobbes wrote this work following a time of carnage and brutality characterized by the English Civil War and the Thirty Years War. Humans, because of their innate nature, were supposedly unable to peacefully govern themselves without the apparatus of the state power represented by the sovereign king. Although debatable as a political philosophy, the notion of the Leviathan has sometimes come to represent the nation state and its many apparatuses of power—courts, legislatures, police, military, bureaucracy, and health and education institutions--as the monarchy has disappeared or become merely constitutional as a symbol. The Leviathan was meant to respond to the threats of war, famine, disease, crime, and pillage. The proliferation of the nation state as the standard for political governance has been the result of the last 400 years.

Geographers, Geoff Mann and Joel Wainwright (2018) have updated the concept with the notion of a "Climate Leviathan" that has to deal with the threats on a much more dangerous scale than the Thirty Years War. All of humanity is affected by climate change in the context of a worldwide atmospheric commons that has surpassed 400 parts per million of carbon dioxide—a situation that could be irreversible. Everybody everywhere as well as future generations are threatened. Particular nation states do not have the sovereign power on their own to solve the problems of climate change, because the issues are far beyond the jurisdictions and boundaries of any sovereignty that each has.

Parallel to the circumstances of the 17th Century what could be the Climate Leviathan of today? Mann and Wainwright see four broad options as being relevant. Two of them are in the capitalist category. The most terrifying is not really a solution at all but the very opposite. Writing at a time when Donald Trump was the American President and the president of Brazil was Jair Bolsonaro, they call it *Climate Behemoth*. In both cases, there was a cavalier denial of even the reality of climate change itself along with frequent boorish mockery and disregard for any kind of environmental concern. These presidents and their regimes ignored or repealed many environmental laws for the protection of biodiversity and previous limits on oil and coal production and logging were aggressively reversed. The very worst excesses of extractive neoliberalism were actively encouraged and bolstered by mindless right-wing populism. A truly dismal circumstance that still finds its application recurring in many parts of the world and, alas, could arise again even in traditionally stable democracies.

The second approach within a capitalistic framework is what, for the time being, they *do* call *Climate Leviathan*, because it does have nominal authority through the United Nations and the vast majority of nations states collectively through several treaties, These include the Kyoto Protocol of 1993, the Copenhagen Accord of 2012, and the Paris Agreement of 2015 and the annual follow-ups of the Conferences of the Parties (COPs) where adjustments and means for achieving carbon reduction targets are discussed. In these cases, 192 nation states are supposedly collectively engaged in reducing carbon emissions so that the global temperature does not arise above 1.5 degrees Centigrade. Quotas are determined, pledged, then sometimes renegotiated, and in far too many cases broken or ignored by major polluters. This is where the most dithering and stalling occurs with the culprits including the USA, China, Russia, Canada, India, Australia, and Saudi Arabia each jockeying to further their positions in maintaining industries that release carbon. Corporate lobbyists—egregiously including those from the petroleum industry--are busy at these meetings helping to delay any tangible action. Mann and Wainwright (2018) lament all this almost meaningless frenzy as a way of preserving the hegemonic hold of capitalism, and any of the positive measures tend to overlap with green growth proposals.

Two other possibilities are considered that could be framed within a broad spectrum of socialism—although extremely different. One, a maximally authoritarian top-down one, is labelled *Climate Mao* after the notorious and brutal deceased Chairman of the Chinese Communist Party—Mao Zedong. Not yet in existence but still latent in the People's Republic of China, such a regime through its draconian powers would be able to compel hundreds of millions to sacrifice personal wealth, obey restrictions, experience harsh rationing, and, through massive physical labor, participate in projects with little to no use of carbon-emitting machinery—all for the sake of mitigating the sources of climate change. Such a response could emerge in a wider regional context of Asia where the power of states and their rulers have maintained traditions of collective obedience and sacrifice. With a growing dominance in geopolitics, China's potential to influence such a drastic, but perhaps even ultimately necessary, response in most parts of the world is not to be discounted.

The final possibility is so far the least visible in its manifestation and yet for which the authors seem to have a fair amount of sympathy. The call it *Climate X* and it would consist of the actions of thousands of peoples' movements, not particularly beholden to the nation state as represented in Climate Leviathan nor to the green growth aspirations of capitalism. Such a situation would involve people taking things into their own hands and developing local economies directly in tune with their particular ecosystems and devising ways to commit their labor without the use of carbon-based fuels. Mann and Wainwright (2018) see Climate X as a place-holding concept for a very wide range of solutions that have not yet fully manifested but are in process and would be taking shape from the bottom-up through multiple forms of participatory democracy based in many cases partly on indigenous traditions. It is similar to the "Post-Development Pluriverse" that Escobar (2015) touts above in the discussion of Degrowth. The notion is also in line with Chapter Thirteen where the importance of grass-roots movements is stressed for actions on issues presented in this book. It also corresponds with the set of solutions outlined in Chapter Fourteen.

Conclusions

Readers would presumably think Climate Behemoth repulsive and Climate Mao would be without much appeal because of all the gains in human rights and personal liberties hard-won over the last several hundred years. Yet I imagine the latter may reluctantly be seen by some as an ultimate possibility because of the appalling lack of action so far. Climate X might have a lot of appeal because of the ideal of bottom-up participatory democracy. Perhaps though, regarding Climate Leviathan, itself, there ultimately will be a final realization that there is no time left for dithering and procrastination. Following Seth Klein's (2020) wartime analogy, the nations of the world might find the equivalent of a "Yalta moment", when the WWII allies became determined and pledged to defeat the NAZIs, and do the same with regard to climate change. Perhaps that is the best we can hope for.

Turning to the three economic agendas, they all agree on one thing—ultimately developing an economy that is not dependent on carbon-based energy and, for the most part, they support some of the same legal and technical means to accomplish that. The degree and intensity of throughput is what seriously divides the approaches of Green Growth versus Degrowth—the latter mandates a massive reduction whereas the former carries on with the desirability of a consumer culture requiring the same emphasis on extraction and an expanding global economy. In my reading of Green New Deal proposals, they seem to be unclear about whether the consumer culture should continue to grow or not.

With regard to government participation, the Green New Deal is explicit on this—it is based on a Neo-Keynesian economic theory whereby governments would provide the majority of financial stimulation and, on a massive scale, direct the whole operation with war-like urgency while generating many new jobs. Obviously, Green Growth advocates still want businesses and the market to control the transformations and allow investors to profit from them. Overlapping with GND ideals, Green Growth advocates do want government subsidies, tax breaks, and other incentives as well as some penalties to guide corporations and investors to a net-zero economy. Degrowth advocates would have to rely very heavily on government funding and robust and binding legislation compelling most sectors of the economy to shrink and make polluters pay dearly through massive fines and taxes. Government funding would be needed for bolstering human care systems such as health, education, social services, and to maintain guaranteed annual basic incomes for everybody.

Regarding the costs of public funding and debt, assertions made by economist Stefanie Kelton (2020), who has served as economic advisor to U.S. Senate Democrats, about Modern Monetary Theory are worth looking at. It is her claim that the only authority that can control the amount and nature of money is the government itself. Governments can release as much money as they choose into their economies to try to realize particular objectives and do not need to be beholden as debtors to the private sector as economists from other schools of thought might assert. (Consider again the fantasy of fractional reserve banking that was discussed in the section on Degrowth—that too was arbitrarily established through government fiat as is the floating value of any currency.) The fundamental warning, though, is that governments should not under normal circumstances release too much money too quickly because it can cause inflation destabilizing economies. Government debt, she claims, is not really the same as household debt—i.e., based on negative balances of assets and expenses. Government "debt" is traditionally covered through treasury bills sold to the private sector and that, she claims, is only done to stimulate investment. In reality and in ultimate terms she asserts that there is no such thing as government debt—a controversial unorthodox suggestion to most of us based on our personal assumptions about household or business balance sheets. In her view, nothing has to be paid back because *the government is its own legal creditor and debtor at the same time, and money itself is now fiat in form based entirely on its issuance and legality by sovereign governments without reserves of gold or silver to back up its value.* Still, it is subject to public and market confidence in determining its floating value and, with heightened globalization, currencies can be volatile to circumstances other than a particular nation's sovereign power.

This all may remain mysterious. Yet during the 2008-2009 Economic Crisis followed by the COVID-19 Pandemic, globally governments pumped astonishing amounts of money into their economies—first to bail out banks and other "too-big-to-fail" financial institutions and more recently to provide their citizens with billions in income supplements because of massive under-employment due to COVID-19 restrictions but also to help keep small businesses afloat. Given the magnitude of the climate crisis, it could be argued even more compellingly that no expense should be spared and such a notion would favor Green New Deals. Two other economists, Yanis Varoufakis (2017) and Kate Raworth (2017) have also devised unorthodox ideas for bringing about the massive amount of financial support to publicly fund projects to deal with climate change and rising social inequality.

Mark Carney (2021) makes the counter claim that only the private financial investment sector can muster the kind of capital to accomplish the necessary transformations. Only briefly and in a off-handed way, though, does he mention the severe damages created by capitalism and refers to them through the euphemism as "externalities". He *does* advocate the need for more equity and attention to social, health, and educational services in the economy but gives next to no detail. He also relies heavily on the ideal of a "mission-oriented capitalism" with corporations rebalancing their agendas to include the aspirations of stakeholders including workers, local communities, and the environment itself rather than just investors and high management officials—the usual winners in the capitalist jackpot. He gives an extended example of an early industrialist,

Josiah Wedgewood a Member of Parliament and crusading abolitionist, of having done that.

Yet as they say—that was then and this is now. I remain skeptical—given the cut-throat competition of corporate capitalism and the demands for lucrative returns by investors and capitalists' proclivity to always look for ways to reduce the costs of production thus continuing to off-shore factories to low wage regions and generate more threats to labor through technological changes of automation. A lack of confidence in "mission-oriented capitalism" tends to be reinforced by the long frequently rapacious history of capitalism that has led to the very problems that we are most concerned about here. That being said, of the three economic strategies outlined in this chapter, Green Growth is the one most likely to be put into operation because of the powerful forces that endorse it.

Regarding green new deals, they tend to be supported by left-leaning political parties and more progressive factions within liberal parties. In the U.S.A., the idea was made most famous by Alexandra Octavia-Cortez, a charismatic Democratic member of the House of Representatives—her resolution for a GND did pass there but it failed passage in the Senate. Yet its ingredients are favored by a margin of 31% by the American public according a poll (https://www.dataforprogress.org/the-green-new-deal-is-popular).The European Parliament has passed a Green New Deal by a substantial margin. In Canada, the New Democratic Party and the Green Party support GNDs, but they are electorally marginal to the Conservative and Liberal Parties that alternate in governing the country. Whether GND bills are passed in their entirety, it is still likely that portions of their ingredients will come to be practiced in many places as the severe impacts of climate change become even more urgent and the public demands an end to the dithering and procrastination.

As mentioned earlier of the three proposals, Degrowth suggestions currently have the least chances of being put into operation. The powers-to-be will not allow them because they clearly upset their vested interests. Similarly, the public in developed countries is far from ready for shrinking economic productivity and possible reductions in its personal wealth. So, the notions are unlikely to gain political support and no political party that has even the remotest chance of participating in governing will advocate it. Then there are the complexities of internationally accomplishing all the adjustments needed to bring about a redistribution in wealth so that the Global South gains in some necessary catch-up growth while shrinking the North's economies.

Yet Degrowth is most compatible with the findings of political ecology. I suspect the majority of political ecologists would endorse it. Accordingly, it is important to keep educating the public and future policy makers on its principles. Also, regarding all three approaches, it is likely that portions of each can be assembled without designating the resulting package as completely representative of any of the three. For instance, from Degrowth--declaring a jubilee (complete forgiveness) or even partial one on Third World debt would go a long way in reducing unfair burdens on developing countries, bring about more social justice, and allow them to more effectively tackle climate change which ultimately affects all countries. Climate-change action anywhere is beneficial to everybody. Legislating against planned obsolescence and generating right-to-repair law along with requiring standardized replaceable parts even among different manufacturers could cut down on manufacturing waste in the North.

Finally with regard to Degrowth, it could be become the necessary contingent reality for many or even all parts of the world sometime in the not-too-distant future—if, as expected, major climate changes lead to crises in economic production along with financial and major blocks to and retreats from globalization. Then with various regions and nations cut off from consistent trade with each other, economies would likely be forced to shrink locally in the context of struggles to survive. Then there is the highly important and unavoidable reality that Bihouix (2014) raised—we are simply going to run out of the resources that have maintained the current high consumption high carbon economy, and there will not be enough of them available to generate an alternative mass-scale green economy. Possibly then with the forced conditions to mold steady-state economies, degrowth/steady state practices could eventually become widespread but with considerable variation in the regional and cultural means through which they are manifested—a pluriverse in Arturo Escobar's (2015) terms.

Such changes would then potentially link economies to their local environments in the way that Nicholas Georgescu-Roegen (1971) an earlier economist with a solid background in physics suggested was the ultimate need for all economic systems. He was the first to contribute to the all-important notion that economies are ultimately limited by the boundaries, laws, and the physical components of matter and energy. Economies, in his terms, are institutions that transform low-entropy matter into goods used by humans but with an inevitable tendency to generate much higher entropy through wastes. They thus need to become reduced in growth before dangerous, depleted, and destructive limits are reached. We are seeing the harsh realities of his insights today. The adaptations that are central to his conclusions are, in the end, mandatory for human survival irrespective of any hopes for a Green New Deal and especially Green Growth. *Degrowth and steady states will have to ultimately prevail or there will be human extinction.* As they say "Nature bats last.". More on this in Chapter Fourteen.

11 Some Solutions in Agriculture

I will avoid here solutions that prolong any approaches promoted by advocates of the industrial, chemical-based, highly mechanized, fossil fuel-driven, monocropped, corporate-dominated agriculture that has been criticized in Part I. These systems have made things much more difficult through so-called "free" trade and neoliberal policies that have led to unnecessarily complicated, grossly unequal, and overly interdependent–commodity/value chains. This has promoted massive, often quite redundant, global exchanges of agricultural commodities based primarily on market and profit considerations—but also led to the unfair agricultural domination in trade considerations and policies by some countries over others. Advantages are primarily intended for the corporation, the investor, and the consumer largely in affluent societies but not the farmer. They include the short-term appeal of low prices for shoppers that hide the real costs that we are passing on to our descendants through damaged environments and social injustice (Patel 2009a). These systems also work against human rights implied by the concept of food sovereignty—for people to produce food and feed themselves nutritiously, within their own regions according to their own preferences, with social justice equally in mind, and most importantly to allow future generations to share in any abundance. The rest of this chapter will explore alternative solutions that emphasize ecological options, thus working with nature, and will be more compatible with the ideology of localization than with globalization.

Agroecology

Agroecology involves the merger of two sciences, agronomy and ecology that for too long have been kept apart because of the high modern domination and attempt to apply an industrial model upon nature. It is coupled with the notion of food sovereignty as promoted by La Via Campesina (LVC) in the scientific, yet participatory, and ecologically directed way of managing agriculture. The Food and Agricultural Organization (FAO) of the United Nations promotes it.

> Agroecology is based on applying ecological concepts and principles to optimize interactions between plants, animals, humans and the environment while taking into consideration the social aspects that need to be addressed

for a sustainable and fair food system. By building synergies, agroecology can support food production and food security and nutrition while restoring the ecosystem services and biodiversity that are essential for sustainable agriculture. Agroecology can play an important role in building resilience and adapting to climate change.

(http://www.fao.org/agroecology/en/)

Agroecology calls for more in the way of a social and political transformation than just a technical one—away from the privileging of agribusiness that touts the technical. The field draws more or less equally from agronomy, local knowledge, ecological knowledge of specific localities, and social science. It relies on integrated trans-disciplinarity and is horizontal rather than top-down in decision-making. Farming and food systems adjust to local ecological circumstances and socio-cultural realities. Sustainability and food sovereignty in the local context, not for profit extraction, are emphasized. Long-term benefits are the goal. Soil fertility is a major concern with importance placed on how to maintain, augment, or restore it based on non-chemical means (Méndez et al. 2016). The approach stresses recycling and the synergistic integration of all aspects of farming and attempts to maximize the diversity that the local environment will allow. In effect, farms are micro-systems in which for the sake of resilience and sustainability, they are considered relatively closed. Energy expenditures especially in the form of fossil fuels are minimized and replaced with animal traction and human labor when feasible. Chemical pesticides are to be entirely avoided and replaced with biological defenses against pests. Insects, microbes, and plant-based repellents and antagonists are used instead.

Diversity is the key word. Agroecologists have been practicing participatory action research (PAR)—a familiar model in applied social science. A much wider range of voices is called upon in trying to reach solutions for local ecological challenges. Those who have normally been marginalized such as farm workers, women, small landholders, and indigenous peoples share in the discussions. Human diversity is recognized as a source of potential innovation beyond the valuable insight such people have derived from their experience in attempting to manage local farming. Diversity in agricultural practices could involve polyculture, more frequent crop rotations, intercropping, and the integration of crops and animals, and genetic variations within species. Regarding diversity in landscape management, that could include such things as buffer zones, contour and strip tillage, and keeping forest fragments intact. It is not just diversity for its own sake, although that in itself has proven to be a general rule of thumb helping to prevent invasion of crop-destroying diseases. The wide variety of species and practices are valuable by combining to provide ecological services such as pest control, nutrient cycling, and pollination (Méndez et al. 2016).

For instance, the traditional *milpa* or "three sisters" system of Central American peasants combined corn, beans, and squash in the same clusters. Corn was the staple producer of carbohydrates and when prepared in the traditional way, soaked in lime, was a source of protein and many significant vitamins. Beans fixed nitrogen in the soil and were an essential source of protein building amino acids. Squash was also a good supplier of protein, essential vitamins, and trace minerals. Its low-lying, runner-based, quick growth served to block out weed competitors.

The member organizations of La Via Campesina all support this approach to the maximum that their local conditions might allow. Cuba, through its national policies, has adopted it. Many universities in Latin America—Mexico, Argentina, Brazil, Cuba, Costa Rica as examples—have research scientists and extension agents doing agroecological work. It is even growing in importance in the homelands of industrial agriculture—North America and the European Union—with important research facilities, for example, at Berkeley, Santa Cruz, the Universities of Vermont, Wyoming, New Hampshire, and Manitoba.

One interesting sideline to be mentioned is that some of the origins began in southern Mexico in the early 1980s as a skeptical response to the Green Revolution (Rosaldo-May 2016). This was in the region of Mayan indigenous peoples. In Mayan languages, there was no word that can be directly translated as "agriculture." The closest equivalent in Mayan languages is *MeyajbilK'aax*, which tellingly translates as *working with nature* or in Nuhuatl the word *Millakayotl*, which means *cultivating with nature*. One of the pioneers is ecologist Stephen Gliessman of the University of California (Santa Cruz) and colleagues (see Gliessman et al. 1981). He recounts a demonstration experiment from the southern state of Tabasco, a hot humid tropical lowland region of previous high productivity through traditional Mayan horticultural techniques. They always considered sustainability on a very long-term basis. The area, though, as a by-product of green revolution policies, had been under an onslaught of attempts at large-scale industrial-styled agriculture such as bananas and sugar cane. After much soil degradation, these cleared forestlands were turned to cattle grazing with even further deterioration. The nutritional status of the local people was declining, and the production expectations of the region fell far short of what the modernist policymakers had expected.

Gliessman and colleagues experimented with a "modular production" form that drew upon traditional Mayan knowledge and practices. It also took advantage of an existing Mayan social system, the *ejido* or village ownership of a commons of land with shared labor and produce. They matched the Mayan diversified system along with early principles of agroecology. The units were 5–15 hectares and managed by several families as a supplement to their other *ejido* activities. The units were surrounded by large shelter belts of secondary regrowth serving as windbreaks, firewood, and building materials and containing wildlife including natural predators and parasites to protect crops. Fruit trees and sources of "germplasm" for further tropical diversity grew there. The center part was the lowest lying and contained fishponds and for use by ducks. Excess moisture drained into this area along with dissolved soil particles, nutriments, and organic materials. The aquatic plants and such sediments accumulated there were used elsewhere for fertilizer. Nearby, raised gardens or traditional *chinampas* with extremely high productivity were created and constantly fertilized with these sediments plus animal fertilizer from the ducks and chickens and pigs kept in pens. These *chinampas* contained a huge variety of annual and perennial crops depending on topography, soil conditions, and drainage. These were grown according to the recommendations of the campesinos. They included traditional polycultural combinations such as corn/beans/squash, cassava/corn/papaya, and fruit trees connected to various types of cover crops, vines, and shrubs. Crop rotations were frequent again following the recommendations of the local campesinos. Altogether seventy-four plant varieties were grown with all-season availability in their combined diversity. Because of the symbiotic biomass recirculation in such systems, yields were extremely high aided by the diversities replicating the natural system. This was all done *without any* use of chemical fertilizers and pesticides. At the time of writing, Gliessman and colleagues (1981) were faced with the challenges of having so much more to learn about tropical ethnobotany in the region and were considering, in the scientific spirit of agroecology, introducing elements of practices from other tropical regions. This, as they point out, would require careful, long-term experimentation before being introduced into the region. So here the importance of localization as the basis for solutions is reinforced.

Given global sustainability issues and environmental degradation caused through industrial agriculture, we desperately need experiments such as these in other parts of the world to start to get beyond chemical addictions in our "wars against bugs" and not rely on chemical fixes to solve soil exhaustion. This applies equally to the Prairies and Great Plains of the United States and Canada. Agroecology is a system of participatory science involving far more than laboratory or field scientists, let alone an all-powerful, dominating corporation.

Miguel Alteri (2009) of UC (Berkeley) is another prominent advocate of agroecology. He stresses the significance of the small-scale farmer, who still feeds the majority of the world. He cites examples from Africa, East Asia, and especially Latin America, where even these regions are targets for industrial agriculture. In Latin America, the small-landholders (average 1.8 hectares) have 35% of the land but produce 41% of domestic consumption in food and more importantly accounting for the majority of staples—51% of maize, 77% of beans, and 61% of potatoes. Their current importance is underscored given that they already operate in an agroecological context based on traditional knowledge, farmer-to-farmer sharing of knowledge, seeds, and other practical agricultural services and products. While still rooted in the past, largely without chemicals, and frequently using integrated polycultures, their systems should be left intact to provide a model for the future. To construct a better pathway, a potential

Some Solutions in Agriculture 167

phase transition that can rival and eventually replace or diminish the current damaging one, it makes sense to reinforce and expand these existing realities.

What kind of logic could contradict such an idea? Possibly that given their non-specialist and older systems, traditional farmers would be unable, if they became the dominant agricultural form, to produce enough food to feed the world at its current and projected population numbers. Agroecologists' response to that claim would be that small farms actually produce more when engaged in polyculture practices taking into total weight per hectare, calories per hectare, protein sources per hectare than the industrial monocrop systems. Polycultural systems are found, for instance, in about 80% of the cultivated areas of West Africa. Farmers there produce grains, fruits, vegetables, and animal products along with fodder for the livestock in the same fields. Alteri suggests that the advantages range from 20% to 60% greater productivity since polycultural systems replace spaces that would otherwise be occupied by weeds and that they reduce the impacts of diseases and insect pests because of the interactions of many species and varieties. Altogether, such cropping systems maximize synergy and efficiency with the mutual interacting uses of water, light, and soil nutrients the way that natural or "wild" systems work.

Farmers then, can also potentially make more profit per output unit. As a projection from intensive farming, he gives examples from the United States where the smallest, two-hectare farms grossed $15,104 and netted $2,902. The largest farms averaged 15,581 hectares by comparison grossed $249 and netted $52 per hectare. Then added to this is the highly significant advantage that such small farmers are able to accomplish this with far less damage to the environment by reducing soil erosion and helping to preserve biodiversity. In the American context, another advantage is that they tend to bypass middlemen and long-value chains, including international ones, where profits are siphoned off by others. Alternatively, they sell directly to customers through farmers' markets, community-supported agricultural networks, and restaurants. They receive better prices because of their reputation for high-quality nutrition and healthier organic means of production. Introducing more of the styles of agronomic know-how added to the already effective traditional small-scale farming in the Global South could solidify and maintain its positions and through the policies of food sovereignty allow families to stay together in strong rural communities.

Returning to the idea of the advantages of diversity, one of the most significant pre-adaptations of the traditional small-scale farmer is that she or he preserves and cultivates landraces—highly varied seed types developed over many generations and passed on from farmers to farmers in largely non-commercial sharing networks. Sometimes peasants, as is the case in potato-growing Peru, can have several scores of varieties of the same crop growing in a single field. These tens of thousands of landraces provide a highly formidable set of resilience modalities against drought, pests, diseases, excess rain, and many other stressors as well as highly varied soil conditions. Regarding local resilience, they can potentially be found on single farms or nearby within the local bioregion. This deeply contrasts with the vulnerability facing the commercially monocrop-committed farmer who specializes in a single GMO variety, for instance, yellow corn, innovated to be planted massively in as much space as possible, used almost exclusively as animal feed, and at the same time is captive to the restrictive dictates of the contracts imposed by his corporate supplier. Then beyond those comparative scenarios, consider that with the huge genetic diversities within the varieties of these landraces, ongoing evolution is constantly happening. With farmer-to-farmer sharing, innovations could be spread more rapidly to farmers undergoing future rapidly changing climatic circumstances. Through scientific collaboration with agronomists committed to agroecology, even more effective new varieties or hybrids could continue to emerge. He stresses that it is important to support policies that protect such zones from GMO contamination that would destroy the highly adaptive and resilient diversity already built into the system. This is so that these systems can continue to provide multiple paths to the future and to even restock the Global North in future circumstances of climate change.

Another asset is that such systems are more resilient to climatic change in comparison to monocrop systems. The diversity allows them some successes along with some failures whereas the monocrop choice is always a big gamble. He exemplifies with African systems of sorghum/peanut/millet in comparison to monocrops. They actually "overyield" when grown mixed together in contrast to when they were grown separately but close in experimental fields with different rates of moisture application. In many cases, traditional farmers practice agroforestry, growing trees and bushes that provide shade reducing temperature, wind velocity, evaporation, overexposure to the sun, and intercept hail and rain from crushing plants. Studies, after Hurricane Mitch hit Central America in 1998, showed that farmers using agroecological approaches fared much better, ending up with 20% to 40% more topsoil thus much less erosion, and fewer economic losses than conventional neighbors.

Currently the dominating ideology pushes for the globalization of production and trade in food commodities with regions and countries pressured to specialize, and have farmers serving export markets rather than local subsistence and supply nearby urban markets. This ties them into long value chains. The longer the chain, the autonomy and agency of the local farmer are proportionally decreased and more power goes to the long line of middlemen. Market pressures are kept up to keep prices low to benefit both corporations and consumers in countries where the commodities are processed and consumed. Such pressures create a trend whereby the farmers have to both expand their holdings and intensify their use of inputs such as patent-protected seed, pesticides, and chemical fertilizers in order, as with North American and European farmers, just to stay in the same position to survive.

Any inputs have to be imported and their prices vary beyond any stable expectation for the farmer. At the time of his writing (2009), Alteri reports a hike of 270% for fertilizers from the previous year, and the prices are often subject to events far away—such as the war in Ukraine. This places enormous pressure on the survival of poor farmers in developing countries and endangers the food security of their regions. Adopting agroecological and food sovereignty policies as alternatives to global industrial agriculture would much better serve these countries and their farmers.

Note: The International Panel of Experts on Sustainable Food has published a set of case studies on successful transitions in agroecology (https://www.ipes-food.org/_img/upload/files/CS2_web.pdf).

Regenerative or Restorative Agriculture

Most of the alternative systems have overlaps with each other and regenerative agriculture is no exception when compared to agroecology. Both emphasize diversity, the rejection of external chemical inputs as much as possible, and promote closed-loop or self-sustaining local systems with low external energy inputs. Regenerative agriculture (RA) is not as holistic, though, not paying as close attention to socio-cultural factors, which are at the forefront of agroecology. What it lacks in holism is more than made up by its obsession with the essential component of agriculture—soil.

The biggest emphasis is placed on its regenerative health through the urgent reduction of erosion, by greatly enhancing organic matter, and expanding soil diversity and quality. Soils should store large amounts of carbon to allow for the maximum growth of plants but also nourish the quantity and diversity of symbiotic microorganisms. RA also counts on the soil itself being a continuing carbon sink to deal with ongoing global greenhouse gas emissions. Advocates stress the importance of preserving large amounts of water through deep, sponge-like root systems and well-aerated soils, especially considering the drought conditions that will accompany much of climate change.

RA advocates want to see soils become as self-contained as possible and more bountiful and sustainable in their capacities to grow crops and feed livestock. Proponents see much less need for artificial fertilizer or pesticide inputs since the diversity generated by healthy soil conditions alone should take care of those needs. The inspiration is nature itself with the observation that ecosystems such as healthy temperate forests or natural grasslands do not need inputs of fertilizers and pesticides. They manage and perpetuate themselves as closed-loop systems stabilizing their fertilities through time.

The most significant aspect of regenerative agriculture advocacy is a determination to favor soil quality through enhancing the extraordinarily complicated realm of microscopic life that already performs vital services for the growth and health of plants. It is vividly described in a recent book about this particular topic in an agricultural context—George Monbiot's (2022) *Regenesis: Feeding the World Without Devouring the World*. In reference to the *rhizosphere,* the narrow belt encompassing the root structures of plants and the rest of this living world, he writes,

> The rhizosphere lies outside the plant, but it is as essential to its health and survival as the plant's own tissues. It is, in effect, the plant's external gut.
>
> Some of the similarities between the rhizosphere and the human gut, where bacteria also live in astonishing numbers, are uncanny. In both systems, the microbes break down organic material into the simpler compounds the plant or the person can absorb. Though there are over 1,000 phyla (major groups) of bacteria, the same four phyla dominate the rhizosphere and the guts of mammals. Perhaps these four bacterial groups have characteristics that make them more prepared than others to cooperate.
>
> Like the human gut, the rhizosphere not only digests food, but also helps to protect plants from disease. Just as the bacteria that live in our guts outcompete and attack invading pathogens, the microbes in the rhizosphere create a defensive ring around the root. Plants feed beneficial bacteria, so that they crowd out pathogenic microbes and fungi.
>
> (Monbiot 2022: 17)

He continues to discuss some of the remarkable symbiosis of plants with soil bacteria. Plants may release, in return for their services, chemicals that kill toxic bacteria threatening benign bacteria that are beneficial to the plants. Plants may then discharge hormones signaling distress to muster the aid of bacteria to attack harmful ones, or plants might reduce their general defenses to bacteria thus promoting an overall predominance of the beneficial ones to do their protective work in overwhelming the toxic ones. In some cases when leaves and stems are attacked by insects or fungi, hormones might be released into the soil that draw bacteria to migrate to counter-attack the invading parasites. Larger creatures may also be alerted by distress signal chemicals—Monbiot (2022) gives the example of nematodes with their sharp beaks attacking underground caterpillars and infesting them with their own eggs leading to the latter bursting open thus destroyed.

Earthworms have long been known to be beneficial to healthy soils. Monbiot (2022) reports that RA healthy soil may contain as many as 8,000 kilometers of their burrows per hectare. They aerate the ground and allow rainwater to trickle through in ways that effectively reach plant roots rather than causing erosion on the surface or forming pools. The worms also reach up onto the surface pulling down leaves and twigs providing nutrients to other soil-supporting organisms and plants. They provide essential minerals through their unique digestive process that involves the uses of tiny rock particles. Worm castings, sometimes in the many tonnes per hectare, brought to the surface further enhance healthy soils.

Regarding this marvelous underground world, the most astonishing component of it may be fungi. Sheldrake (2020) refers to fungi as a distinct biological kingdom, neither plant nor animal, with a possible unique form of intelligence. So far that is only dimly comprehended by science but hinted at by remarkable systems of fungal chemical communication. Fungi are, according to him, the primary originators of soils in the first place by breaking up rocks and along with bacteria continue that role today. Digesting wood and other large bits of organic matter is a highly significant contribution in making carbon available to plants and other biota. Their unique capacity for digesting lignin and cellulose as well as weathering rocks is unsurpassed.

Fungi, unlike plants, cannot rely on photosynthesis as a food source—they mostly have to envelop sources like rotting wood to gain it. Yet this lack of directly generating its food is where the remarkable symbiosis with plants begins. Plants will provide carbohydrates for a very wide range of reciprocal exchanges. With their thin threads known as hyphae, mycorrhizal fungi explore much larger and deeper areas than root hairs can, and the exchanges occur between the root hairs and hyphae. They deliver water, nitrogen, phosphorus, magnesium, zinc, iron, and copper to the plants, create nitrogen from decomposing vegetation, and sometimes help in the transformation of it into protein when there are excesses of it. In all these services to the plants, fungi seem to communicate amongst plant and tree roots, should the latter be present, to distribute these items often roughly according to need.

Endophytic fungi can exist inside the plant's leaves, stems, and flowers and they provide similar functions by drawing nutrients into those regions and generating hormones that assist the growth processes. Both endophytic and mycorrhizal fungi, similar to bacteria, can perform protective functions—killing insects and larvae, generating antibiotics, and converting the corpses to ingredients that feed plants and the rest of the underground micro-biotic realm (Sheldrake 2020; Nouh et al. 2020; Kaviya et al. 2019; van Genuchten 2022, Brown 2015).

Yet in this complex underground world, different species of the same realms of bacteria, fungi, and nematodes can also be the sources of disease and destruction for crops. Once again, this generates the parallel of the rhizosphere that Monbiot (2022) draws with the human gut where bacteria can be essential, protective, neutral, or toxic. He suggests another similarity through the term *dysbiosis,*

> Similarly, in the last few years agricultural scientists have discovered that plants seem to be less capable of fighting off attacks by certain pathogens when they grow in damaged soils with a low diversity of microbes. When the soil has been harmed by too much fertilizer, by pesticides or fungicides, excessive ploughing or crushing by heavy machinery, their cry for help is more likely to be exploited by parasites and pests. In both cases a dysbiosis is caused. This is a medical term, meaning the collapse of our gut communities. But it could be applied to the unravelling of any ecosystem.

An interesting line of research suggests that soils with a rich and well-balanced microbiome suppress pathogenic bacteria that cause disease in people, making the transmission of human diseases through food less likely. Our health depends on ways that are obvious and ways that are not on the health of the soil.

(Monbiot 2022: 20)

Also, key to RA is the idea of having as much carbon as possible in the soil and making soil a highly significant carbon sink in light of the climate change crisis. The unusual *biochar solution* (Bates 2010; Hawken 2017) has been offered as a possible answer. It would involve baking massive amounts of waste biomass, similar to the way charcoal is manufactured, and depositing them in the soils. The idea for this came from a most unusual source—archeological evidence from remote regions of the Amazon tropical forest that were analyzed by a soil scientist.

Amazonian soils tend to be shallow, yellow, and acidic, and when trees and other brush are cleared and burned for the purpose of creating gardens, farms, or soy plantations, any biomass intended to generate carbon and fertility in the soil only lasts a short time. Tropical conditions create rapid rates of decay and then heavy rains leach nutrients from the shallow topsoil beyond the level where roots can reach them. These carbon-deficient plots are only fertile for a few years or have to be pumped with large amounts of fertilizers when, for instance, turned into soy plantations.

In contrast, there were pockets of *terra preta* (black earth in Portuguese) soils that were augmented by the biochar process and accounting for about 10% of those located beside the Amazon drainage system. They were black, highly fertile, up to 6.5 feet deep, and in some cases supported productivity for over five hundred years. This was at a time before the age of European colonization when the Indigenous populations were quite large and occasionally urbanized. Virtually all the waste, fish, bone, kitchen crumbs, broken pottery, manure, and so forth were baked without exposure to air—the *pyrolysis* process and buried underground with a thin layer of cover soil. This introduced massive amounts of carbon into the ground. Then, though tragically after flourishing for hundreds of years, well over 90% of the Indigenous population died by disease or were murdered through the holocaust caused by Europeans.

Supporters for reviving the practice point out its capacity to provide carbon for growing plants plus supporting the extensive and complicated networks of fungi, bacteria, insects such as ants, nematodes, earthworms, and burrowing animals that serve and enhance the growth and protection of the crops. Biochar is highly porous, and its surface area is incredibly large within a small area. The pores set up niches for tiny organisms serving plants, and the carbon in the biochar allows for their proliferation. Biochar attracts nutrients—for instance, a negative electrical charge draws in calcium and potassium ions that are positively charged. Among other things, it can reduce excessive soil acidity created by nitrogen fertilizers.

According to Hawken (2017), there is still much research needed to be done, though, because it does not work in all soil situations. Where it does work, crop yields increase an average of 15%. Soils that are highly degraded and acidic respond very well, and these are found in zones of the world that are having issues of food insecurity. A discovered beneficial side effect is that biochar allows plants to absorb nitrate fertilizers in greater quantities, thus allowing their applications to be significantly diminished. Reducing their application that has generated much wasteful run-off that harms aquatic and other systems is significant.

Regarding biomass on the surface, when decomposing releases methane and carbon into the atmosphere. According to Hawken (2017), under biochar soil circumstances, much of that carbon would instead be stored underground—rendering it stable and fertile for hundreds of years, thus slowing down the normal carbon release cycle.

There is still much to be researched on biochar including its manufacturing process to consider it as *the* solution. Among the issues is the cost. George Monbiot (2022) reports that the cheapest source that he found in the United Kingdom was £1300 a metric tonne. Altogether, the process may be more useful in tropical soils than in temperate ones, but that is also a plus for the former.

An always central part of RA is *zero-tillage* and as a more common practice than biochar infusion—it also helps to maintain and enhance carbon content. Zero-tillage contrasts with traditional plowing and discing, where the previous year's stubble and subsequent weed growth are plowed. Zero-tillage prohibits opening the ground surface. One reason is that the exposed earth is always subject to erosion from rains, wind, and down-slope gravity. The biggest concern about plowing for RA is that it allows for huge amounts of carbon released from the soil into the air. Hawken (2017) reports that about 50% of the carbon previously present in global soils has been lost by tillage over the last few centuries.

The saved carbon encourages microbial growth that serves the crops—roots become deeper and water is preserved. The well-nourished plants are much more pest resistant, and fertility is enhanced, reducing or eliminating pesticides and fertilizers. Successful competition against weeds is accomplished by planting the fields with a wide variety of cash cover crops such as clover, vetch, or rye. Previously, plowing would be used to destroy weeds by tearing up their roots. The use of the cover plants is quite extensive—they also include alfalfa, mung beans, cow beans, lava beans, hemp, kale, mustard, turnips, radishes, sunflowers, sorghum, and others. Plants such as these bring particular positive attributes such as fixing nitrogen, shading out weed growth, and encouraging phosphorus, zinc, and calcium to become available. Farmers using RA are learning how to maximize the benefits by rotating among these varieties.

Another dimension to soil regeneration has to do with methods of pasturing ruminant animals such as beef and dairy cattle. Currently in conventional agriculture, they are now raised through two basic methods—first, confined feedlots, which represent the worst circumstances of environmental and health concerns. The other common method has been to place the animals in a pasture in the spring and allow them to randomly graze there until the fall—but that can lead to overgrazing and deteriorating pastures. The cattle tend to randomly graze haphazardly all over the place. Nutrient reserves in the roots deteriorate and the soils can reach a state of exhaustion. That method leads to "soil compaction, excessive water run-off, gully erosion, persistent weed invasion, and lack of cover to support biodiversity" (Franzluebbers et al. 2012: 1). With this kind of overgrazing, there are well over a billion acres of severely degraded pasturelands in the world (Hawken 2017). Considering that there are advocates for a completely plant-based diet, one conclusion has been to remove all the herbivores from these lands to permit the soils to recover and then transform them into croplands for growing grain exclusively for human use (Monbiot 2022).

Counter advocates such as Hawken respond that grazing in ways patterned after those of wild ruminants are good for the soils and biodiversity. Ruminants such as elk, buffalo, sheep, goats, and antelope as herd animals, being instinctively quite fearful of predators, graze in tight groups in small areas for only a few days. They munch the grasses down to the crowns rather than the roots; defecate and urinate in a confined area; churn up the soil with their hooves; and in the process deposit their wastes as an equivalent of fertilizer. They then move on to another area leaving the former grazing area for a long time. This allows it to go through a healthy process to regenerate with grasses reemerging from the established, deep, perennial roots, remaining highly fertile with lots of carbon, with capacities to support much biodiversity, and preserve and manage water, and prevent erosion. Those wild grasslands, such as found in the East African savannahs and remaining North American long grass prairies where buffalo resided, consist of a great variety of perennial grasses and are extremely healthy and noted for their capacities to preserve water and store carbon.

So, the recommended imitation of these conditions is known as *managed paddock grazing*. The pastureland is divided up into many fenced paddocks where the cattle are rotated frequently and engage in "mob feeding" in each restricted area for a couple of days. They do not return to any particular paddock for at least a month in warm humid conditions and at least a year in cool dry ones, thus allowing grasses and soils to regenerate.

Farmers who have used the method have reported all sorts of benefits—in some cases they can increase herd size by as much as 200%; it encourages the return of deeply rooted perennial native grasses; it ends the need and expenses of sowing domesticated grasses using fertilizers, pesticides, and diesel fuel; and it increases capacities to soak up much more rain rather than creating erosion-causing run-off. The cattle change their behavior moving quickly from paddock to paddock and also eating the weeds rather than just the grass, and the weeds are found to have high protein contents. Beyond these benefits, the pastures sequester much carbon and thus have a positive impact regarding climate change.

Regenerative agriculture has much to offer in research, especially in soil science where we are just at the mere frontiers of knowledge, cannot but benefit all kinds of agriculture.

Note: The National Resources Defense Council has an interesting website on restorative or regenerative experiences in an American context (https://www.nrdc.org/stories/regenerative-agriculture-101).

Natural Systems Agriculture and Perennial Grains

Eighty percent of our food consists of cereal crops, oil seeds, and legumes, and, depending on the country, they take up 50–80% of cropland (Pimental and Burgess 2013). They are primarily raised through industrial agriculture as monocultural systems requiring huge amounts of energy, water, fertilizers, and pesticides. The staple crops of wheat, maize, and rice dominate constituting from 35% to 50% of our average daily caloric intake (Soto-Gomez and Perez-Rodrigues 2022; Glover and Reganold 2010). Overwhelmingly they are annual crops that require yearly replanting from seeds produced in previous years.

Why annuals? Back about 10,000 years ago when humans started to domesticate grains from wild grasses, post-glacial environmental conditions were unstable, and annual grasses were opportunistic to those circumstances. With much larger and more plentiful seeds, they were able to quickly, with shorter growing seasons, abundantly occupy and dominate disturbed zones and temporary floodplains. Requiring less time to grow, their roots were much shallower than perennial grasses. The annuals, as adaptable opportunists, were, through their seeds, able to move quickly from their temporary zones to yet newer ones. In contrast, perennial grasses occupied their territories for a long time thus placing much more emphasis on the growth of stalks, leaves, and, especially, deep roots, rather than on larger and more numerous seed heads since their need for rapid reproduction was much less. In nature, perennials dominate, annuals primarily take hold as first responders after catastrophic conditions, and then the perennials establish themselves spreading deep roots and a multitude of interconnected symbiotic and competitive species.

For early agricultural experimenters, most presumably women, the annuals proved superior as potential food sources. From year to year, they would select from those annuals that had the most desirable characteristics regarding flavor, abundance, utility, and ease of harvest. Through many generations, these annuals became distinctly separated from their wild cousins—converting them into the many successive varieties of our modern wheat, rice, and corn. Perennial grasses were left behind, although many of them also had potential for useful domestication. The whole complex of agriculture became largely dedicated to the cultivation of annual grain crops, which require constant and time-consuming care. They tend to have shorter growing seasons, thus earlier cultivation and quicker adaptations to new zones, such as the limited growing periods of the Canadian prairies. You could even say that these grains have also domesticated us—we certainly have been the reason for their spreading so widely (Glover and Reganold 2010; Manning 2004) in yet another phase transition with a significant path dependency.

Richard Manning, though, in his book *Against the Grain: How Agriculture Has Hijacked Civilization* (2004), suggests these adaptations and subsequent ones involving livestock led to the worst cultural, social, and environmental consequences that can be attributed to humanity. They resulted in huge population increases, social inequalities including slavery, predatory states, genocidal wars, many new diseases and other health problems, and major environmental disruptions.

George Monbiot (2022) suggests that, through its practices, the annual grains systems emulate the catastrophe-like conditions where it had its beginnings. Because of the ancient and standard procedure of plowing before sowing, it has been the source of huge amounts of soil erosion, degradation, and loss of fertility. Since the roots of annuals are so shallow, soils also may not be effectively bind and serve as a source of water conservation. Standard process weeding stimulates circumstances parallel to the original condition of annuals, where at their very beginnings, by swamping an ecozone, they then have fewer competitors. Much carbon also has been lost because of erosion through plowing and exposure, and the heavy use of pesticides has destroyed huge numbers of animals and plant life that are symbiotic with healthy plants and soil. The expenses of modern, industrial, monocrop agriculture have led to a continuous reduction of farmers and farm families, impoverishing rural communities, and generating very large social inequalities and conflicts. Energy expenditures and climate change consequences have been huge because the system of annual grain production requires many passovers of fields with diesel-burning heavy machines that also tend to compact soils. Because of pressures on farmers to grow lucrative cash crops to serve their household needs, ensure some stability, and deal with cost-squeeze pressures, there has been a tendency to repeat the growing of the same grain, legume, or oilseed year after year rather than apply environmentally useful rotations of longer fallowing with cover crops. This leads to over-dependence on artificial fertilizers. We are running out of phosphates, and excessive nitrogen with its more frequent toxic run-offs has led to serious environmental consequences.

So, in light of all this, a number of researchers and advocates for alternate systems (such as Soto-Gomez and Perez-Rodrigues 2022; Glover and Reganold 2010; Monbiot 2022; Cox 2009; Jackson 2002) have promoted a major changeover through the development of perennial grains. While annuals have had an over 10,000-year head start, current genetic knowledge and hybridization procedures allow for the fairly rapid development of new species and varieties. The biggest issue is that perennials store much more of their carbon in their roots and stems than annuals. The challenge then for agricultural productivity is to get much larger seeds on multiple seeded heads with hulls less easy to shatter so that they do not spill before harvesting. Significant progress has been made but much more has to be done.

The advantages of perennial grains are extensive. From other perennial examples such as alfalfa, they are effective in preserving topsoil—with the carbon in the soil increasing 50% to 100%. Because they have longer growing seasons and very much longer roots, sometimes in the many feet rather than inches, they retain water and nitrates much more effectively. With greater carbon content, they will help mitigate climate change because they will reduce the need for inputs. Because of their deep root structures and sponge-like conditions, more water will be stored helping to adapt against warming and drought. They can be used to rehabilitate marginal and previously degraded lands thus giving farmers more cash crop opportunities. Farmers will not have to perform annual seed-bed preparations and planting. If a farmer has been sick one year and not able to cultivate fields, he or she will be able to return next year with both soil and water having been protected (Glover and Raganold 2010).

Since they do not require annual planting, they lower input costs, raising profit margins and helping to increase the family's well-being or even survival. Along with drought resistance and longer-term stability, higher aboveground biomass allows for more abundant mulching materials, biofuels, and even animal grazing. A more carbon-rich and carbon-storing rhizosphere becomes more complicated and interconnected, allowing for more nutrients and exchanges through micro-biotic action (Gomez and Rodriguez 2022), and energy costs are greatly reduced (Pimental and Burgess 2013). Varied perennial ecosystems also promote much higher rates of biodiversity both below and above ground. Also, land does not need to be taken out of production through fallowing or cover crops since perennials also serve the same functions as the latter but often for lucrative and necessary cash crops (Monbiot 2022).

Results from experiments with perennial grains have been encouraging. A perennial rice, PR 23, has been developed in China and gets yield results equal to annuals even sometimes exceeding them and in two yearly harvests. The hybridization involved a common Chinese annual rice with a wild perennial grass of the same genus from Africa. As of 2022, it has already gone through 12 harvests without reseeding. Seventy thousand hectares were at this time being grown, and farmers were anxious to adopt it even with supplies still limited. The variety adapts well in binding soils in China's crowded landscape, especially in sloping conditions thus aiding in preventing erosion. There is a rural labor shortage because of young people moving to the cities and industrial jobs. Farmers could be potentially relieved of much of the labor intensity in rice growing most especially during the annual transplanting periods. Monbiot (2022) reports that

Yunnan University is now in the process of hybridizing PR 23 with other varieties that have different uses and taste requirements. Drought-resistant perennial rice is being developed for upland conditions that are much more subject to major water shortages (Glover and Reganold 2010). This general situation for upland farming constitutes about 14 million hectares or 11% of the world's rice growing areas and farmers there tend to be poor and precarious through drought and erosion (Pimental and Burgess 2013). Such initiatives when fully developed will be a hugely important contribution in the context of climate change to the most populous and rice-eating regions of the world.

For North America, the leading innovators for developing perennial grains and oilseeds were Wes Jackson and his colleagues. As a non-profit research organization, Jackson established the Land Institute at Salina, Kansas, in 1976 where they began the search for useful perennial grains. Jackson importantly notes in what he calls *natural systems agriculture,*

> The never plowed native prairie serves as our teacher. Nature's prairie features a diversity of species, nearly all of whom are perennial. The perennial root system is the underlying strength of the prairie ecosystem. This ecosystem, thus, maintains its own health, runs on the sun's energy, recycles nutrients, and at no expense to the planet or people. Another consideration, wherever we look, from the Canadian prairies to Texas, from the state of Washington in the west to Ohio in the east, roughly 2,000 miles in both directions, wherever there's prairie, four functional groups are found: warm-season grasses, cool-season grasses, legumes, and composites. Other species are present, but these groups are featured. Different species thrive in dry years, others in wet ones. Some provide fertility by fixing atmospheric nitrogen. Some tolerate shade, others require direct sunlight. Some repel insect predators. Some do better on poor, rocky soils while others need deep rich soil. Diversity provides the system with built-in resilience to changes and cycles in climate, water, insects and pests, grazers, and other natural disturbances.
>
> (Jackson 2002: 115)

Jackson writes that their research requirements from that reality are to replicate diversity and perennialism and somehow keep the diversity implied in the four functional ranges of grasses in producing grains suitable for human use. Maintaining the *structure* of the prairie is essential to retaining the *functions* required in that ecosystem. Research is directed to requiring that seed yields be increased in these perennials so that there is no cost to other parts of the plant, especially the roots. It is important to work out a polycultural system that outperforms monocultural annuals. Can the new systems adequately manage pests? Can the perennial polyculture supply all of its nitrogen needs? With regard to this approach, he suggests that his "natural systems" strategy can be applied worldwide *but* with the understanding that it be "devoted to developing species and mixtures appropriate to specific environments" (Jackson 2002: 115). That contrasts with transgenic and Green Revolution strategies of plant breeding that emphasize uniformity across regions. Jackson instead embraces localism and diversity.

Some other requirements for plant breeders in these alternative systems include high yield, a synchronicity so that all the grain can be harvested at the same time, and that there be a retention of seeds on the plant until harvest. Attributes that ease sowing and harvesting are also sought. Very precise and painstaking investigations are required through the examination of thousands of possible species or varieties to be chosen and plants that could be hybridized with successful ones. After several generations of successive planting though, they may reach a dead end (Monbiot 2022).

At the time of Jackson's 2002 writing, they were making progress at the Land Institute on three of the four functional groups on the native prairies. These included two warm season grasses—a dwarf maize and a sorghum, a legume, and an oil-producing sunflower. The fourth required grass function—a cool season grass—appeared later in the 2000s (DeHaan et al. 2018). It was first labeled as an intermediate wheatgrass, although later renamed Dernza by the Land Institute to help promote its distribution. With it, the Land Institute had taken over a project first conceived in 1987 by the Rodale Research Center of Pennsylvania, where regenerative agriculture was first conceived. Beginning in 2002, plant breeders at the Land Institute annually started growing plots of the wheatgrass selecting the seeds from the most prolific plants in each of six breeding cycles. Seeds are still smaller than wheat and yields are only about 26% of those of wheat. Attempts are being made to hybridize it with annual wheats, but so far perennialism is lost—attempts will continue. The Land Institute is cooperating with colleges of agriculture at the Universities of Minnesota, Wisconsin, and Manitoba. It is grown on a small scale by about 400 farmers who have experimental plots while growing conventional grains (Lanker et al. 2019). Corporations such as General Mills and Patagonia Products sponsor it. Although the market is tiny so far, beer, bread, other baking products, and a breakfast cereal by General Mills have been successfully produced. It can also be used directly as animal fodder or it can be applied as post-harvest, stubble-field grazing. It is best grown intercropped with other perennials such as clover, thus maintaining both the structure and function of prairie ecosystems as Jackson's (2002) criteria.

This is all very promising but there is still much work to be done—increasing yield, further hybridization attempts, dealing with plant diseases, exploring dwarf versions that reduce stem sizes, and diffusing its use through to more farmers, corporations, and customers, beyond the earliest adopters. Monbiot (2022) estimates that if all goes well, it would take about 30 more years to catch up with the status of annual wheat. Wheat, though, has had a head start of 10,000 years—here we are talking about 70 years for Kernza.

Note: For more information on perennial grains and the Land Institute see https://landinstitute.org/our-work/perennial-crops/.

Some Other Considerations and Conclusions

There is much more than can be discussed here. *Permaculture* (Holmgren 2002) has many overlapping similarities with agroecology, regenerative agriculture, and natural systems agriculture. Involving biomimicry, it is place-based, first through detailed observation of local environmental relations. Then by means of a set of 12 principles, which generate highly planned and intensively integrated procedures, relatively small-scale operations are established or reshaped with a maximum of productivity in plant and animal diversity and generating considerable closed-system recycling. Permaculture seems most useful for small farms and new homesteads.

Wasted food looms large as a food security issue and represents a source of much global inequality because of its over-representation in affluent countries. According to Hawken (2017), hunger daily affects about 800 million people, and the food which we squander represents 4.4 gigatonnes of carbon dioxide released every year—about 8% of humanity's greenhouse gas emissions. Besides more than just fair global distribution, much needs to be done to reform market and retail procedures to prevent the discarding of unsold food or withdrawal on the basis of artificial "best before" dates.

We, especially in the affluent regions, are eating far too much animal protein both from over-fished oceans and from the massively scaled practices of raising domesticated land animals. In the latter case, too much of our cropland is devoted to raising feed for them. The environmental consequences are huge. So, plant-based substitutes should take over a larger proportion of our diet (Hawken 2017).

Questions—will our future continue to be based on ever-increasing off-farm migration and urbanization, and therefore a possible intensification of the need for large-scale farming with fewer farmers to provide the necessary staple crops? Or will there be a return to rural-based, localized, or regional marketing, small-scale farming in a future where globalization has collapsed or diminished, and there is a need to adapt to climate change and local conditions?

George Monbiot (2022), although an enthusiast for regenerative agriculture and perennials, pessimistically claims that massive urbanization is probably here to stay and the need for staple grains, especially wheat, for city dwellers will continue to expand for quite a while. This means, in his reluctant opinion, that there will be a need for the kind of large-scale industrialized farming that we see in the wheat-producing areas of the United States, Canada, Australia, Argentina, Russia, and Ukraine. As imports, these needs will also continue for many parts of especially the impoverished world as a result of dependencies generated by globalization. He does not see localization or urban farming in any significant way being able to counter such realities.

In contrast, both Jason Bradford (2019), *The Future Is Rural*, and Chris Smaje (2020), *A Small Farm Future*, take the position opposite to Monbiot. The only way for humans to be resilient and survive the eventual upcoming disruptions of climate change will be through low-carbon, localized systems that are generally frugal and based on effective small-scale farming and rural communities—in effect the same sort of settlement patterns as our ancestors experienced but with much superior scientific knowledge.

Returning to the materials presented in this chapter, it is easy to summarize since they are all of a kind. We probably could legitimately label them, along probably with permaculture, under the same banner—agroecology—since they are all ecological and deal with agriculture.

Healthy soil, free from erosion, including micro-biotic diversity and quantity, elevated carbon content and storage capabilities, along with water preservation, is fundamental—all aided by zero-tillage. External inputs including energy are to be avoided as much as possible, focusing on closed loop systems where most materials are recycled. Polycultural and perennial crop solutions should be attempted whenever feasible. Having solutions that are localized makes the most sense, since, in order to maximize productivity and sustainability, it is prudent, environmentally, to "go with the flow" as to what Nature provides rather than to impose standardized industrialization upon it. Diversity in crops and animals should be preserved through land races and heritage animals discovered by several thousand generations of innovative peasants and farmers. They should form the basis of new hybrids or retrievals of them by agronomists to meet climate change challenges. Diversity, also, in socio-cultural context among other things, promotes social justice but also generates much better solutions with those with invaluable local experience working equally with committed agricultural scientists.

There are concrete examples of all of this through Cuba's (see Rosset and Benjamin 1994) national policies on agriculture and the Landless Workers Movement in Brazil (see Patel 2007). Yet given the global hegemony of the corporate food regime, and the biases of most corporate-linked colleges of agriculture, it will take a concerted effort by the kind of political agroecology of mass movements as recommended by Gonsalez de Molina and colleagues (2020) to gain significant traction for these fundamental reforms. In the meantime, we can legitimately hope that some of the specific practices can be assembled and diffused even into conventional forms of agriculture.

Whatever the case, any significant transformation will demand much more patience and understanding of the very fine details of ecological relationships in highly localized circumstances rather than the modernized world views of short-term domination, convenience, industrial ease of effort, and massified uniformity.

12 Renewable Energy Solutions

There are extremely well-qualified and respected PhD-level physicists and engineers who have shown that there are technical solutions to climate change that involve renewable non-carbon emitting sources of energy—sometimes in a net-zero context. The net zero-connotation refers to the fact that there might be a few remaining supplementary fuel sources such as biofuels made from domesticated sources or waste vegetation. They are renewable themselves and will serve as sources of carbon capture when regrown thus roughly drawing back the carbon that they released. More controversial and not shared by any advocates that I cite is the belief that carbon capture technologies of sequestration can aid in the reduction of greenhouse gases thus permitting some fossil fuels to continue to be burned.

The most senior of these physicists and engineers advocating is Amory B. Lovins, a well-respected energy "guru," advisor to businesses and governments, and director of the Rocky Mountain Institute. An advocate since the 1970s, his proposals can be found in *Reinventing Fire: Bold Business Solutions for the New Energy Era* (Lovins and Rocky Mountain Institute 2011) containing technical solutions along with carefully integrated designs ensuring energy savings. His most noted emphasis is that of "negawatts" (Lovins 1990, 1996) or energy *not expended*—huge savings through efficiencies in power production. Florescent replacing incandescent lighting is a familiar example.

Another prominent advocate is Mark Jacobson, Professor of Engineering at Stanford University, who, along with Mark Delucci (2009), wrote an influential article in the *Scientific American* on how the whole world could by 2030 achieve energy sufficiency through renewables. Subsequently, he has written two books also in encyclopedic detail that likewise provide multiple solutions. One (Jacobson 2020) is highly technical, probably meant as a textbook for engineering students, and the other (Jacobson 2023) titled *No Miracles Needed: How Today's Technology Can Save Our Climate and Clean Our Air* following the same format is meant for the educated layperson. He refers to the new regime as a "WWS world" based on wind, water, and solar sources of energy.

A third major source is Saul Griffith, an Australian-American, PhD engineer, inventor, and entrepreneur of renewable sources of energy, who has founded multiple companies in this field. He is also known for devising intricate charts

DOI: 10.4324/9781003495673-15

of American energy flows, identifying locations for savings and opportunities to make easy transitions to renewables. He maintains that American 30,000,000 jobs could be created through such transitions. Like Lovins and Jacobson, Griffith's (2021) *Electrify: An Optimist's Playbook for Our Clean Energy Future* underscores the huge opportunities for business profits and general growth and prosperity. These postures all belie the notion coming from extreme right-wing sources that advocacy for renewable energy is a left-wing plot. Instead, they all fit comfortably within mainstream American values of "can do," entrepreneurial expectations. Other sources based on quality technical expertise (Elliot 2020; Peake 2021; Hossain and Petrovic 2021; Hawken ed. 2017) have been informative for putting together this summary of renewable energy.

They all present convincing arguments as to how a world could be transformed to be completely reliant on non-carbon emitting sources of energy taking into account projected global population increases and needs for increased energy needs, especially in the developing world. They *all* oppose the use of nuclear power in achieving this transformation. This is because of the enormous costs of nuclear power, the dangers of military nuclear proliferation, terrorism threats, and the reality that, by the time enough reactors are in place, the damages done by our fossil fuel regime will have been even more extensive and will have led us into the most extreme of climate change possibilities. Problems of nuclear waste remain unsolved; uranium sources will eventually run out; and future generations, already excessively burdened with our negative legacies will be left with the responsibility for dangerous nuclear ruins, exorbitant costs of decommissioning, and having to deal with highly contaminated areas—all without any benefits from the power.

Considering advantages from a revolutionary shift to renewable energy, Mark Jacobson (2023) points to the millions (7,000,000 annually) of lives that could be saved by the elimination of the major sources of air pollution all caused by the burning of fossil fuels. Lovins and Rocky Mountain Institute (2011) raise several other highly significant advantages. The ultimate sources of energy will be largely free in stark contrast to trillions of dollars spent on exploration, production, and the processing of fossil fuels. Their much lower costs could relieve the average citizen from some of the current high costs of living. For the United States, they suggest an extremely important bonus—in that huge military expenses could be greatly reduced since so much of them are devoted to the United States getting unhindered access to petroleum from foreign sources. (Although realistic cynicism about the ultimate nature of the United States' military-industrial complex might temper that optimism.) For all of humanity, the rapid switching to non-fossil fuel sources of energy is essential for any possibility of halting further life-threatening consequences of the climate crisis.

The core authors (Lovins and Rocky Mountain Institute 2011; Jacobson 2023; Griffith 2021) make intricate recommendations for a multitude of essential domains—all types of transportation; manufacturing including heavy industries that require extremely high temperatures; buildings—domestic, office, commercial, plant, and public; electrical generation and storage; and innovative redesigns of the electrical grid. All three reject underground carbon sequestration as energy-wasting schemes as expensive boondoggles that will not ultimately have any significant effect but have been promoted by green growth advocates as a way to achieve "net-zero goals" while still using coal and petroleum as fuels.

Stephen Peake (2021) points out the sun delivers in 115 minutes (less than two hours) the equivalent of all the energy consumed by the world's economy in a given year. Jacobson (2023) suggests another positive bonus is that such a revolutionary shift could end many energy insecurities since all countries at the very least have access to the sun and wind and river systems for run-of-the-river or unfettered micro-hydro. Further Hossain and Petrovic (2021) list a series of countries—Denmark (69.4%), Brazil (75%), Austria (80%), Norway (98.5%), Costa Rica (99%), Paraguay (100%), and Iceland (100%)—that have already accomplished 100% renewable transitions in electricity production or are tantalizingly close. Other countries that are making significant steps in such directions are Sweden, Germany, New Zealand, (64%) Morocco, Scotland (97% in 2020), Uruguay, and China the largest developer of renewable facilities while still the world's largest CO_2 emitter (https://www.climatecouncil.org.au/11-countries-leading-the-charge-on-renewable-energy/).

There are no logical justifications to delay these transitions especially since many suggest that more jobs will be gained than lost and those in fossil fuel-dependent ones can be retrained. Yet fossil fuel companies have successfully slowed down this essential transition because of their lust for squeezing as much profit as they can from their remaining deposits and they also fear the losses of stranded assets such as the many billions of dollars of infrastructures left in the Alberta tar sands (Dembicki 2022; Mayer 2017). They could switch their considerable capital assets to renewables, and besides as Joseph Schumpeter (1942) and more contemporary economists and capitalists have suggested enormous benefits gained for the economy and society through these kinds of "creative destruction." There are other kinds of ideological and cultural barriers to such an ultimately species-saving transition.

The rest of the chapter will provide brief summaries of the multiple sources of renewable energy; ways to achieve much energy savings; comment on some of the solutions to perplexing problems, such as energy storage, keeping constant flows of energy especially during peak periods of demand; and providing without fossil fuels extremely high temperatures needed for manufacturing, as well as other suggestions such as high degrees of localization and diversification, and the all-important factors of redesigning grids and power plants. While being highly optimistic in that we already have the solutions for the energy domain, highly important caveats and barriers to their fruition will be discussed in the conclusions.

Sources of Renewable Energy

Solar

Hawken (ed. 2017) points out the literal significance of the term solar farm with panels harvesting photons from the sun and energizing electrons creating electrical currents—light to voltage. Solar can now compete on its own with all other forms of power generation—not requiring any fuels, cheaper than most, and will continue to grow more so. Hawken (ed. 2017: 9) not only refers to it as a solution but as a "revolution." Still, because of its intermittent nature and with its noon-hour peak a few hours before the highest daily demand, it is important to continue to develop cheaper storage and battery solutions along with smart-grid coordination with steadier sources such as hydro, geothermal, and tidal.

Solar photovoltaics (PV) involve no moving parts and convert solar radiation directly into electricity. They consist of solid shapes, are modular, reliable, require little maintenance, produce no emissions, make no noises, and last a long time. This is all based on the discovery in 1839 that certain materials when exposed to light will emit electrons—the photovoltaic effect. The first uses of solar panels on land began in the late 1950s and the first "farms" in the early 1980s. Development has been rapid with major growth occurring around 2010 due to a great reduction in the cost of materials and through temporary government subsidies. The semiconductors in the panels can be constructed from many materials beyond silicon and the less the crystalline structures of the materials, the more the capacity of photon-generated electrons to easily flow through the materials and the less the loss of energy (Hossain and Petrovic, 2021). Major strides continue to be made. Added to the potential of PV is the use of mirrors and lenses that can be used to concentrate solar energy up to 1000 times more thus reducing the amount of silicon, which had become a major part of PV expense.

Concentrated solar power (CSP) is another solar technology—instead of converting photons, it uses solar heat as the equivalent of a fuel that boils water into steam to run electric generators. Tilted mirrors are arranged so they follow the sun's rays during the day, concentrating its heat at a particular point within a tower containing molten salt, oil, or water, and then operating a steam generator to create electricity. A key part of the technology involves molten salt that stores high temperatures for up to eight hours and that means the generators can keep going when light rays diminish and demand rises. Desert and very dry regions have predominated for this technology, with the western United States and Spain being its initial centers, but with Morocco now surpassing them in importance. Other potential favorable areas include the Near East, Mexico, Chile, China, Australia, and the Kalahari Desert in Southern Africa (Hawken ed. 2017).

Wind

Wind power is on a par with solar in promising a non-carbon energy future and as such has been around for a while in the form of windmills (Peake 2021). Today it is found in the form of wind turbines that in 2017 globally numbered 314,000 and supplied 3.4% of global electricity and, as one example, powered ten million homes in Spain alone. Denmark, the significant pioneer, supplied 40% of its demand in 2017. The United States is especially favored with the winds blowing in Kansas, Texas, and North Dakota, estimated to have such resources to meet all of the country's electricity needs (Hawken ed. 2017). Technically, wind resources could supply up to six times the current global production of energy (Peake 2021). Wind has some advantages over the sun in that it blows at night, is strongest during the winter when energy needs are high, and, along with solar energy, is popular in the public imagination (Elliot 2020). Wind farms are advantageous because they only take up about 1% of the land that they occupy, leaving lots of room for growing crops or raising cattle, with the farmers receiving rents from the leases and revenue from their crops or livestock. Wind farms can be constructed in less than a year, bringing relatively quick returns on investments—compare that to nuclear reactors that can take up to two decades to begin to operate. Wind power is getting cheaper in 2017—2.9 cents per kilowatt-hour as compared to natural gas at 3.8 and solar 5.7 (Hawken ed. 2017).

Wind tower blades are similar to airplane wings made of composite materials—carbon-fiber-reinforced plastics and glass and are designed to be flexible; the larger the blade, the more power produced—increasing that area also reduces the cost of electricity. Like photovoltaics, wind farms require a transformer and connection to a nearby grid—converting direct to alternating current so that the power can be sent greater distances (Peake 2021). There are two types of propeller designs—vertical and horizontal. The verticals have upward-facing spindles that are capable of responding when the wind shifts direction. The horizontal ones, with one, two, or three propellers, are more common because of the potential space covered by the large blades and the towers have to be widely distributed to effectively capture the wind. The sizes and thus the power of the turbines have been gradually increasing. The largest in 2021 is near Rotterdam in the Netherlands using one blade at 107 meters producing 12 megawatts (Hossain and Petrovic, 2021).

Taking advantage of high ground makes sense to catch more prevailing winds, as do offshore locations where winds are more constant with fewer land barriers. However, a disadvantage is the wear and tear, thus earlier long-term deterioration of the equipment and more difficulties in maintenance. Offshore wind farms have been built fairly close to shore, with the North Sea as a primary location. One example is Hornsea One, lying off the Yorkshire coast, with 174 X 7 megawatt turbines on 100-meter towers, providing enough energy to serve one million homes. Other developments in the works include the building of towers that can reach 100–200 meters in depth and floating turbines, both expanding the areas where offshore wind farms can be located (Peake 2021).

There are disadvantages—the principal one being that wind is intermittent and rather than being compact, they do require a large amount of space between each tower. On land that can be overcome by combining the operations with other kinds of enterprises such as farming. With regard to intermittency, Hossain and Petrovic (2021) recommend the construction of storage batteries on site and the expanded use of excess energy for the manufacture of hydrogen fuels (see below).

Hydro-electric Including Micro or Run-of-the-River Hydro

Given the revelations of Chapter 8 concerning large-scale hydro dam projects, it would be hard to seriously advocate here for their expansion. Yet as it stands now hydro provides over 60% of the world's renewable energy—about 4.5 terawatts globally or about one-sixth of the world's electricity (https://ourworldindata.org/renewable-energy) and about 7% of our total energy production (https://ourworldindata.org/sources-global-energy) so in spite of its sometimes negative social and environmental impacts, it is an important source of non-carbon emitting energy. Relatively speaking, there are some countries that have been able to construct hydro projects without as much human and environmental damage—Norway (95% of its electric power), Iceland (76%), Canada (61% with some huge caveats), and Sweden among them (Hossain and Petrovic, 2021). To what extent they should be extended remains a compelling question only to be answered by experts with great care and after intricate impact analyses. These projects are expensive, and other forms of renewables (sun and wind) are becoming much cheaper.

Small-scale hydro based on electric generators placed in rivers without impeding them can be arranged to produce as much as 30 megawatts, enough to power 6,000 high energy-consuming American households. The installation process is relatively inexpensive, and compared to wind or solar, it is not subject to the vagaries of weather with almost no negative impact. The flow does need to be fairly constant throughout the year, and the overall downstream drop must be at least three meters to allow for enough pressure to run the turbines. Some systems have free-flowing diversion channels running beside the river to the generator then further on rejoin the downstream flow. For isolated communities, there are large advantages in that there is no need for expensive fuels such as diesel to run electric generators (Hossain and Petrovic, 2021).

Hawken (ed. 2017) sees great potential especially in regions like the Himalayas where rapidly melting snowfields provide "hotbeds" of in-stream activity that could rapidly and cheaply propel rural development. There are underwater analogues to wind energy with blades being rotated as water moves past. No diversion or storage is required, and upkeep is minimal. The technologies are still

188 Solutions

relatively new and rare so much yet could come of their development. Hawken (ed. 2017) indicates that large parts of the continental United States are promising with Alaska, the Pacific Northwest, and the Missouri, Ohio, and Missouri Rivers as potential zones. Also, they have a role in reinforcing localization along with appropriate smaller-scale or convivial technologies.

Ocean Power (Wave and Tidal)

The energy from the oceans is, so far, along with hydrogen and its fuel cells (see below), among the least developed and far behind wind and solar where research and development is concerned. That energy comes from three possible sources—wave energy driven by the wind, tidal energy based on the daily gravity pulls of the moon, and thermal energy drawn from the ocean. The last, while having some potential, is the least of the three and only significant in southern regions.

Peake (2021) outlines the technologies: wave energy converters—energy from waves collected by bobbing snake- and duck-shaped devices; tidal range schemes where water is made to flow through turbines and capture energy from rising and falling tides; tidal and ocean current turbines using large propellers on the seabed or suspended below the surface making use of currents including tides. Another type of wave converter is known as the oscillating water column approach—examples are located along coasts with waves striking enclosed columns of air with the resulting pulsating upward and downward movements of the waves moving electric turbines (Hossain and Petrovic, 2021). According to Peake (2021), the IPCC estimates that there is about the equivalent of 15 terawatts in the oceans, and that is considerable, although the specifics about its distribution are still vague.

Regarding waves, size matters—the power of waves being proportional to their height. Peake (2021) illustrates this—a three-meter wave is nine times as powerful as a one-meter wave. Ninety-five percent of a wave's energy is found in the surface layer. Wave power, like wind power, is intermittent and therefore has little potential for contributing to any baseload security. Yet the European Union is developing plans whereby wave power could supply 10% of its energy by 2050. Wave power would be best managed on coasts where westerly winds converge, as in the east coasts of the British Isles or the region around Cape Town, South Africa where the South Pacific and Atlantic converge, the east coasts of Brazil and Madagascar, the southeast coast of Canada, the northwest coasts of Spain and Portugal, and generally in the latitudes between 40° and 60° (Hossain and Petrovic, 2021).

Tidal assemblages considering regular ocean currents as well as the two daily tidal flows are more reliable and could contribute to regional base-loads. Again, according to Peake (2021), the IPCC estimated that there was a range up to 3 terawatts available, yet citing another study regarding the 28 best sites that was estimated in particular as only 360 gigawatts but still ten times the massive Three Gorges Hydro Complex in China. There are currently two major sites—one in France and one in South Korea. Technologies for the sites have to be reversible to manage the in- and out-flows with propellers or tall crossflowing vertical spindles. Besides coastal tides, there are a few regular ocean current flows that seem promising—such as the Gulf Stream flowing between the Bahamas and Florida.

One region of special note regarding wave, tidal, and wind is the Orkney Archipelago just north of Scotland with 20,000 people. Laura Watts (2018) won the Rachal Carson Prize for her account of the innovation and community solidarity of the people there in her *Energy at the End of the World: An Orkney Island Saga*. While describing the remarkable energy enterprises there, she provides a sociocultural portrayal of a resilient people facing the storms and multiple and complicated tides of the islands with the winds almost constant. Through their efforts with enterprising and experimental companies, much is being developed, and virtually all homes have small wind turbines. A much higher proportion of the population there than anywhere in the United Kingdom drives all-electric vehicles. Altogether the renewable enterprises provide 115% of the local energy needs. Much more could be produced and to serve the needs of mainland communities, but the underwater cables are too limited in voltage to increase transfers. Interestingly, one of the companies uses some of the excess in a one-half megawatt electrolyzer to produce hydrogen. Although not yet allowed for regular fuel, at the experimental stage, docked ferries are powered at night by hydrogen fuel cells (see below). Watts (2018) points out that the experience of the Orkneys, while being highly innovative and ahead of most of the world, points to a theme reinforced in this book—that of intricate and diversified localized solutions—they will be required in the challenging regimes of the future, and that certainly applies to the details of any renewable energy mixtures. Much more optimistic than some authors, Paul Hawken (ed. 2017) suggests governments need to promote research and provide incentives for innovation and implementation of such projects in this domain because, while still in its infancy, this sector could serve up to 25% of the United States' electricity needs, Australia 30%, and Scotland 70%.

Geothermal Energy

Geothermal sources can be used to generate electricity and for heating of water and buildings both individually and in districts. The heat for all these functions is transferred from the inner core of the earth through the convection of liquified rocks upwards through the inner, outer core and mantle and then closer to the earth's surface. Conduction is then common with molecules and atoms colliding through the earth's crust of solid rock, thus then through vibrations. In current contexts, it can be tapped as sources in depths up to a few kilometers.

The zones that are most suitable for this source are along the borders of certain lithospheric plates of the earth's crust. One example, the "Ring of Fire" a massive such formation and one most suitable for such energy sources is found in an arc swinging northward from Australia northward, past Asia to the Arctic, southward along Western North America, and then South America. In North America, the Cascade Region from British Columbia to Northern California is notable with the geysers of Yellowstone National Park as an example of naturally occurring surface phenomena. Other places can be found in Central America, The Philippines, and Japan. Iceland, along another lithospheric plate boundary, is another prime example.

The best sites involve an aquifer that is capped above and below by impermeable rock. High heat from below creates steam in the capped aquifer. Fissures are drilled or found naturally from the surface; the steam is used to generate electricity as well as piped for heating buildings and delivering hot water. Another drilled or natural fissure provides inputs of rain and surface water to keep the aquifer replenished. In the majority of cases, though, rather than being able to rely on existing aquifers depend upon dry heat with holes drilled to the source of geothermal heat, water poured in, steam arises to operate electric generators, then is cooled to be inserted downward again to recycle the process (Hossain and Petrovic, 2021).

Hawken (ed. 2017) reveals that 39 countries could garner 100% of their electricity needs through geothermal energy. Kenya gets half of its through plants in the Great Rift Valley, Iceland one-third, and El Salvador and The Philippines one quarter. He suggests that in order to facilitate the huge potentials, governments need to get more directly involved because of the costs of any initial drilling can be quite high but the later on-going benefits considerable.

Bioenergy

Bioenergy was the earliest form of energy used by humans for heating, cooking, making pottery and bricks, and smelting metals. Its use is still quite considerable and probably will be for a long time yet. Wood, dried animal dung, and charcoal are deeply embedded in the domestic economies of peoples in the developing world.

For renewables, bioenergy sources come from recently living organic matter than fossilized ones and their carbon is, at least conceptually, net zero since it is recycled with the equivalent more-or-less captured equally through regrowth of the same plants. Biomass can be found in wood, crops, waste, and derivatives such as fuels and are collectively seen as part of a transition to a complete non-carbon economy or WWS regime. Nonetheless, Jacobson (2023) is opposed to their use because when used as fuels for transportation usually mixed with petroleum, they are still contributing to climate change emissions and air pollution and when they are used for heating and electricity because they are sometimes used in the context of carbon sequestering that consumes as much as 25% of the energy produced in the process of capture.

Unlike Jacobson, Hawken (ed. 2017) sees biomass, in spite of its flaws, as a necessary "bridge" to a WWS world because it can generate electricity upon demand unlike solar and wind power and is largely recyclable. The carbon would naturally return to the atmosphere through decaying plant material anyway. Biomass should draw upon appropriate sources such as waste products and carefully selected cash crops. Corn and sorghum should not be used because they deplete ground water, contribute to erosion, and use energy in expensive equipment, pesticides, and fertilizers. Hardy perennial crops of short rotation such as switchgrass could be used as well as ones that can be effectively grown on marginal lands not suitable for food used by humans. Woody crops including poplar, eucalyptus, and shrub willow can be cut low to the ground and then grow and be reharvested quickly. Waste from agriculture and the milling of wood can be used. Biomass currently fuels about 2% of global electricity production and among the Baltic nations of Sweden, Finland, and Latvia—20% to 30%. Biomass production needs to be tightly regulated so that large companies do not make market decisions to expand the niche too extensively thus using up highly valuable farmland needed to feed humans (Hawken ed. 2017).

Another biomass methodology (Hawken ed. 2017) is methane digestion. As organic matter through decomposition and microbial action waste from animal digestion, agriculture, and any organic activity produces methane. It is highly combustible and contributes as a greenhouse gas up to 34 times stronger than CO_2, although not as lasting in impact. But being placed in a sealed metal digester, the gas can be used domestically for heating and cooking and the solid waste provides a concentrated highly nutrient-laden fertilizer. This is a very promising alternative to the burning of wood, dung, or coal in the developing regions of Africa and Asia, and on an industrial level can be used for the production of heat and electricity. Municipal landfills are as yet still a major untapped source of methane produced electrical energy.

Hydrogen

Mark Jacobson (2023) is big on hydrogen as a fuel for the future—mainly in solving certain issues in transportation—heavy long-distance trucks and large airplanes flying trans-continentally. Hydrogen is extremely abundant in nature but is always compounded with other atoms as in H_2O, so therefore must be synthesized through electrolysis—electrically splitting the hydrogen off of the water. That requires energy, so as another bonus beyond its abundance; he points out that any electricity that is overabundant through variable solar and wind can be diverted to electrolysis, thus serving as a kind of storage system. Saul Griffith (2021) is not keen on seeing it only as a narrow niche fuel for the future and as rather expensive and as wasteful in requiring large amounts of energy to produce energy.

Back to its advantages as Jacobson (2023) and others have noted, compared to fossil fuels, its waste after burning is not carbon dioxide but water. With reference to the Hindenburg tragedy of the 1930s, which drastically inhibited further consideration of hydrogen as a fuel, he maintains that the blimp did not burn so many people to death because of the hydrogen itself, but because lightning struck a highly flammable coating on the blimp's skin. Hydrogen supposedly only burns vertically, not both vertically and horizontally as with jet fuel or diesel, making the latter two much more dangerous besides their huge contributions to the climate crisis.

Regarding its production—there are two ways of manufacturing it. One, sometimes referred to as "gray" or "blue," involves a synthesis from a carbon-emitting medium—natural gas—being called blue if it is accompanied by carbon capture. The other is labeled as green and involves the electrolysis mentioned above and is the only one allowed in Jacobson's WWS formulation as it eliminates drilling, transportation, processing, pollution, and the release of carbon capture technologies. The hydrogen is normally compressed or liquefied and stored for later use.

Its use in a 100% non-carbon world as advocated by Jacobson would be in fuel cells to run electric motors in a range of vehicles, for heat and electricity production, and for heavy industry such as in the production of steel. He points to a company in Sweden that produced steel through the oxidation of hydrogen that was made renewably mainly through wind power. Given the potential for driving large vehicles including aircraft, there are still some issues with regard to storage—hydrogen is way less dense than gasoline and jet fuels and 13% of its energy has to be used in its liquidation, so tanks have to be big—1.8 to 4.8 times larger to go the same distance as jet fuel (Jacobson 2023). Regarding their use in cars, they are much less efficient than all-electrical battery-run vehicles. Yet with heavier vehicles such as long-distance trucks they overtake battery-fueled electric cars and there will be many advantages with regard to air travel. Elliot (2020) predicts that the lowering costs of green hydrogen through hydrolysis will overtake gray and blue through methane syntheses before 2035, thus scaling up its significance in the non-carbon economy.

Note: The United Nations Sustainable Water and Energy Solutions Network has a series of renewable case studies that may be of interest to the reader (https://www.un.org/en/water-energy-network/page/case-study).

Transportation

Electric vehicles (EVs), drawing power from batteries, produce zero tailpipe emissions. While they still may have to draw electrical energy from carbon-emitting power plants, the amount of pollution accumulated from them is 15 to 30 times less than from road traffic generated by conventional combustion-run vehicles. Mining, processing, and manufacturing associated with EVs also cannot escape carbon emissions, but in a future WWS world that too all will eventually be replaced by clean renewable sources. Battery-run EVs have a further advantage in that, even with braking, this kinetic motion actually restores energy into the battery. EVs also are more efficient than internal combustion ones—only 17–20% of the energy in the latter is used to move the vehicle and a large part of the energy is used to drive the weight of the vehicle rather than the passenger and its cargo. The rest of the energy is wasted as heat. With electric vehicles 64% to 89% of the electricity runs the vehicle and, as they consistently get lighter, more is available for the passenger and/or cargo (Jacobson 2023).

Besides the benefits of the electric powertrain, Lovins and Rocky Mountain Institute (2011: 24) see "the integrative whole-system design optimized for ultralight materials, particularly advanced composites" creating "what we will call the 'Revolutionary' auto—the key to getting autos off oil by 2050." The automobiles keep getting lighter and powertrains smaller and simpler. Weight savings are the key with each design cycle and with the eventual ability to place electric motors in each wheel thus eliminating the need for clutches, transmissions, driveshafts, and axles.

Without sacrificing strength, long carbon-fiber composites provide for continuing advances in achieving lower weights. They are encased with plastic resin to create materials tougher and stiffer than steel and a third less in weight while not rusting or fatiguing over time. The composite materials can reduce by ten times the number of parts that constitute an autobody, lowering production costs by lessening ports and robots to complete the assembly.

Trains and trams: the first electric passenger train to run was developed in Berlin in 1879. Soon after electric tram-lines using a third rail or overhead wires became global running in urban settings followed in big cities. Yet there were major regressions with fossil fuel run vehicles taking over—diesel-run locomotives, trucks, buses, torn-up tram tracks, and the consumer-driven obsession with personal vehicles run on gasoline.

Another reversal is now occurring—there have been major recent expansions of electric-run trains—especially in Asia. China has over 100,000 kilometers of such systems, India 46,000, and 85% of both its passenger and freight is electric. Japan is at almost 100% in this regard, and Germany and Turkey are similarly expanding. Tractors and heavy equipment are slowly being electrified in Europe with the major inhibition being a shortage of charging stations. Regarding marine transportation, a significant number of Baltic Sea ferries have been electrified. Jacobson also notes innovations underway in short-haul electric aviation. For him, the significant breakthrough will come with hydrogen fuel cells for long-haul air travel. Such aircraft could have the same thrust-to-weight and range as today's combustion aircraft (Jacobson 2023). He sees similar advances for long-range shipping and trains. As I write this section, a company in Quebec has been taking passengers on test runs with a hydrogen fuel-cell train (https://www.cbc.ca/news/science/hydrogen-train-quebec-city-1.6888891).

Lovins and the Rocky Mountain Institute (2011) raise the question the overuse of the personal vehicle, which contributes greatly to greenhouse gas emissions. Solutions lie in maximizing the extent and reliability of public transportation, more emphasis on car sharing and borrowing, and constructing workplaces and shopping facilities that are closer to homes. Suburban sprawl has been a major impediment to tackling the problem.

Similar concerns are raised regarding long-distance trucking of goods. He offers a number of steps that could be taken to improve the appallingly low average mileage costs of diesel transportation (4.5 to 6.4 miles-per gallon exacerbated by the practice of long idling) through similar advances being made in electric cars. Much more could be gained through "intermodal freight" that involves sea and rail transport (much cheaper and more efficient) with short-haul trucking to final destinations. Road congestion would be ameliorated and in the long term, there would be no need to transport fossil fuel, especially coal, by rail that now takes up such much cargo space on both trains and ships.

Buildings

Regarding buildings, the fundamentals are providing electricity plus heating and cooling. The electricity function should in part be taken care of through the common use of solar panels on the roofs, storage batteries, the use of electric vehicles being charged but when fully charged being available for immediate use in the building if needed. They would be attached to whatever grid connections are locally suitable for delivery of renewable energy.

Jacobson (2023) lists multiple approaches to keeping them warm or cool. One of these is district heating and cooling where air or water is heated at a central location and then loops through the buildings and rooms on its links distributing the heat through radiators then return to the station to be heated again. The reverse occurs with cold water during hot weather. What he calls "third-generation district heating" involves highly insulated pipes running underground allowing hot temperatures but still below 100 degrees Celsius thus saving energy from the need to boil water first and, if needed, using electrical energy from renewable sources for that. Water in tanks can be heated by the sun and beyond heating in a building that water can be used for its other domestic purposes. Heat pumps can be used for cooling and heating water as well as buildings. Another method is freezing water into ice when electricity is cheap or in excess from renewable production, and then running water through pipes that are covered with the ice when cooling is needed. Several other methods raised by Jacobson (2023) involve storage of hot or cold water underground to be used as seasonally needed for temperature control all without the need for fossil fuels.

All three of our main sources (Jacobson 2023; Griffith 2021; Lovins and Rocky Mountain Institute 2011) enthusiastically tout heat pumps that require no fossil fuels such as natural gas—just electricity to operate them, which, in turn, would eventually be produced by renewable sources. Either way, they reduce or eliminate the need for fuel costs and reduce greenhouse gas emissions. Situated next to the buildings, they extract heat or cold from the outside air and deliver it inside according to seasonal needs. The principles use liquid refrigerant and fans in reverse directions evaporating or condensing according to need—evaporating refrigerants and drawing hot air out of the rooms and in cases of cold air the condensed refrigerant draws heat from the coils and then redistributes it inside. Some forms of heat pumps make use of pipes running underground and draw from heat and cold in a mode of geothermal energy extraction.

With reference to the United States in 2011, Lovins and Rocky Mountain Institute (2011) point out that it contained 120 million buildings. In their pursuit of energy savings, they reveal that this sector consumes 42% of the nation's energy, 72% of its electricity, 34% of its natural gas, and that people spend the significant majority (87%) of their lives inside. American buildings, in their opinion, have been energy hogs. The remedy first lies in insulation. The shells in their forms, structures, and functions determine to the largest extent their energy uses in lighting, heating, and cooling. Poorly insulated and leaky ones add considerably to energy costs. Aligning them in the right direction with regard to the daily course of the sun and properly managing their openings can enhance their temperature controls and lighting without further cost. Spoiled by abundant and relatively cheap electricity, many people do not turn their devices off (especially computers) when not in use. Smart metering along with sensors helps with that in the long run. Some new technologies such as "smart windows" darken in response to heat; "thermochromic" windows using a liquid-crystal coating will vary the amount of heat entering a building—increasing solar heat by as much as five times during the winter and saving about 30% of average heating bills. Altogether, as with sectors such as transportation, they call for integrative design instead of having each sub-sector of construction and interior facilities designed separately. Given the large stocks of buildings already in existence, retrofitting policies and government incentives are imperative.

Industry

There is a huge demand for energy in manufacturing, and, in the United States' case, three-fifths of that is directly from fossil fuels and also hydrocarbons have become necessary for feedstocks such as plastics, nylons, and chemicals. Much of the energy goes into heating things, sometimes at very high temperatures as in smelting metals, making steel, rubber, concrete, plastics, and glass, down to tiny dabs of solder on circuit boards. It is also used to power all the machines such as robots, drill presses, and conveyor belts engaged in the many processes of manufacturing (Lovins and Rocky Mountain Institute 2011).

For the kinds of transformations needed, Jacobson (2023) sees hope for replacing fossil fuels in high temperature functions such as the smelting of metals. One of them will be hydrogen as in his example above of steel-making in Sweden. He further demonstrates how electricity can be used directly for that and many other processes that require prodigious amounts of heat. One significant technology is the arc furnace—spherical-shaped with a retractable roof and a heat-resistant floor for collecting molten metal. The process involves three graphite rods that can be raised or lowered into the furnace. Scrap steel or other metals are placed in the furnace, and the rods are lowered. A high electrical current is passed across the positive and the negative rod forming an arc with the anode creating an exceptionally bright light and raises such high temperatures that the smelting is accomplished. Jacobson (2023) also describes other furnaces—induction, resistance, dialectic heating, and electric beam heaters—that can perform similar functions without fossil fuels. Heat pumps can be used to produce steam. Lovins and Rocky Mountain Institute (2011) suggest a whole series of significant energy savings so that manufacturing industries can be more energy-efficient in the transitions to an all-electrical industrial regime based on renewables.

Storing Renewable Energy

In conventional non-renewable electric systems, storage was never particularly of concern. As Hawken (ed. 2017) points out humans have devised clever storages for almost all our products but those for the "commodity" of electricity are still underdeveloped. Essentially, electricity keeps flowing until it is used as centralized power plants produce it based on calculations of demand and when they have surpluses, they send them further down integrated grids to distant utilities. In a WWS regime, there will be a need for ingenuity applied to storage issues since it is true that the sun is not shining all the time and neither is the wind blowing.

To some extent there can be the equivalent of continual available baseload capacities in hydroelectric systems, both traditional and run-of-the river, geothermal and tidal systems but they will be limited by regional circumstances. So, batteries are essential and lots of them as Griffith (2021) suggests. Chemical batteries are great for storing energy on a short-term basis—for daily variations in availability—say for up to a week but not as effective for the long term as in the case of preparing for a long winter. Their costs have been falling rapidly and more cuts are likely to come. Lithium-ion batteries were US$1,000 per kilowatt-hour (KWh) in 2010 and expected to have fallen to $75 by 2024. They are good for about 1,000 recharges each, and there are needs for research into methods to extend those cycles. So, in the future halving the costs and doubling the recharge cycles will add a lot to WWS carbon-free future security. Griffith (2021) also looks to the parked electric vehicle as battery storage. Given an average 80 KWh battery and with say 250 million such vehicles that would represent 20 terawatt-hours—an enormous quantity and stored when not in use that could be made available as electricity for the grid. Jacobson (2023) points to concentrated solar production where heat stored in molten salt allows for storage to continue running electric generators for as much as 15 hours.

Regarding battery storage and mainstream grids, Hawken (ed. 2017) reveals that Los Angeles was in the process of replacing its peak demand gas power plant with 18,000 batteries to be stored with electricity from wind and solar power. He discusses another phenomenon of note. In the past, centralized power plants dominated the consumer by setting the process and delivering the power. In the future, with more distributed systems, individual homes with solar panels or plugged-in cars and other small-scale energy systems such as wind farms can send excesses upwards into a larger grid, delivering the power to more distant places in need.

Another method, although quite minor in comparison to the others, is the use of flywheels. Here a wheel or spinning disk of heavy material continues to spin after basic energy-drive movements in a machine that has stopped. Kinetic energy stored from previous motions allows for continued operations. Some have been constructed from carbon fiber materials and can achieve up to 60,000 rotations a minute. They are good for storing energy for short periods (several hours) and require little maintenance.

Longer-term storage can be considered through other methods. Pumped hydro was mentioned in Chapter 8. It is good for short and medium storage but given the possibility of large reservoirs can be longer-term. Again, that involves the pumping upwards of water into large storage facilities built into a valley or as part of existing hydro facilities (Griffith 2021). Compressed air is another method. The air is pumped into a cavern, salt dome, or enclosed vessel. Then, as required is released in an expansion turbine that generates the electricity. Although reminding one of "Rube Goldberg machines," similar in principle to pumped hydro-storage are gravitational storages with solid masses. One method involves using a crane that stacks solid blocks of concrete through excess energy—then when needed the procedure is reversed, and the motions of the crane generate the needed electricity. Another version involves a train carrying rocks or concrete blocks being powered up a hill and then reversed downhill when needed.

Finally, manufactured hydrogen can also be considered the equivalent of a battery or storage system in that a way to produce it would be to use excess renewable energy that has been produced in other ways.

Bringing It Altogether through Redesigning the Grid with Major Diversifications and Nested Localization

Anthropologist Gretchen Bakke (2016) wrote *The Grid: Fraying Wires between Americans and Our Energy Future*. Its holistic advantage soon becomes evident in this superb example of the growing subfield of energy anthropology. Her analyses and informants include multiple stakeholders—engineers, utility company, and government officials, consumers, and many more. From her multiple-angled perspectives, she can perceive difficulties and flaws that others might not notice. For instance, while planners might see the possibility of 250 million electric vehicles being plugged into the grid providing a kind of giant battery ensuring baseload security, she questions whether the car owners themselves would tend to see themselves as cheerful, compliant contributors to this greater public good. Or in viewing a situation where affluent California home-owners had installed solar panels on their houses and had subsequently profited from feed-in tariffs through their surpluses, she points out that in order to maintain its profits, the local utility company charged higher electricity rates than before to much less affluent renters who were equivalent to captive consumers. One of the cultural inhibitions to major changes that she notes with her fellow Americans is their essential dependence not only on petroleum but on electricity in their expectation of easy access and uninterrupted flow. She even suggests a German term that would translate as "seeky" in characterizing their unconsciously close-to-addictive reliance on it.

She points to the electrical grid as humanity's greatest architectural achievement and as probably the highlight of the 20th century. Yet now it is massively inefficient and in danger of being completely worn out. It has been encrusted by years of ownership by utility monopolies, top-down corporate control, overreliance on extremely large centralized power plants in turn highly dependent on coal or nuclear with frequently costly break-downs, and major regional electrical blackouts that take days to uncover the causes. The companies that control it all largely practiced a command-and-control ethos that has been highly resistant to change.

Fortunately, as Bakke (2016, 2021) and all our other major sources recognize the emergence of diversified local systems characterized variously as micro-grids, granular modular grids, and connected islands of grids. All of these authors support resiliency from the bottom-up that break or threaten corporate monopolies. Ownership is varied, from highly competitive start-up companies of various sizes and capital, to local cooperatives, to individual ownership at the home level. From passive consumers to changes in some circumstances lead to Lovins and RMI (2011) characterizing them as "prosumers" showing agency in making decisions that may profit themselves by selling to the grid from solar panels.

These systems are quite diversified, adding significantly to the range of solutions where previously companies would rely on single sources—coal, nuclear,

Renewable Energy Solutions 199

hydro. They are quite local in the range of solutions chosen that are appropriate to the regions in question. Added to the possibility of a successful transition to renewable power is the existence of smart grids, which can better make adjustments at the household and regional level when shortages or surpluses recommend shifts in consumption. Bakke (2016) gives the examples of automatically allowing temperatures in freezers to rise just a few degrees when there are shortages of electricity or allowing loaded washing machines to take advantage of surpluses at low prices at unorthodox hours. Smart metering and sensors can be made available to govern much larger grid systems to manage flows among the nested local versions in larger connections parallel to the older grids.

Mark Jacobson, the acknowledged technical expert in the field, has succinctly put together a series of steps to create a full WWS electric world, so it is best to quote him in full.

A. Transition transportation to run on batteries and hydrogen fuel cells, where the hydrogen is produced from electricity.
B. Transition all building air and water heat so that it is provided with heat pumps, direct solar and geothermal heat, and/or district heat. Also ensure that district heat is produced by electric heat pumps, direct solar heat, and/or direct geothermal heat. Use district heat immediately or store it in water tanks or underground. The types of underground storage include boreholes, water pits, and aquifers.
C. Transition air conditioning in buildings to run on the same heat pumps as those used for heating in buildings, or run a district cooling loop that is part of fourth generation or higher district heating system. The district cooling system should include cold storage in water tanks and/or ice.
D. Electrify high, medium, and low temperature heat generation needed for industry with electric arc furnaces, induction furnaces, resistance furnaces, dielectric heaters, electron beam heaters, C.S.P. steam, and/or heat pump steam.
E. Transition all remaining combustion appliances, machines, and processes to electric ones including moving to electric induction cookers, electric fireplaces, electric leaf blowers, electric lawnmowers, and electric chain-saws, for example.
F. Build wind farms in cold regions because wind output and heat demand both increase with lower temperatures, and the wind electricity can meet the heat demand.
G. Interconnect, through the transmission system, geographically dispersed wind, wave, and solar resources to turn some of their variable supply into steadier supply while reducing overall transmission requirement.
H. Build both wind and solar together, since wind and solar are complementary in nature.

I. Increase energy efficiency in buildings by using LED lights, installing energy-efficient appliances, reducing air leaks in buildings, and increasing insulation in walls and windows.
J. Improve passive heating and cooling in buildings by using thermal mass, ventilated facades, window blinds and films, and night ventilation.
K. Provide all electricity with onshore and offshore wind, solar PV on rooftops and in power plants, CSP, geothermal power, tidal power, wave power, and hydroelectric power.
L. Store WWS electricity in CSP storage, pumped hydropower storage, stationary batteries, vehicle batteries, flywheels, compressed air storage, and/or gravitational storage with solid masses.
M. Build more short- and long-distance high-voltage AC transmission, long-distance high-voltage DC transmission, and AC distribution lines to interconnect WWS supply and demand centers (Jacobson 2023: 288, 289).

In this imagined WWS world, demand overall should decrease because there will be no need to use energy to mine fossil fuels and uranium, nor energy required to process and deliver them. Again, other than the costs of operating the equipment there would be no more fuel costs. Plus, energy efficiencies will continue to rise in importance and government and corporate policies will encourage less energy use. With regard to supply, it is important to provide wind and solar outlets in the same region and it is important to interconnect geographically disbursed sources of wind, solar, wave, tidal, and geothermal sources to common transmission grids and that will quite significantly lead to the smoothing out of power supplies. Related to that with regard to winds being intermittent, Jacobson suggests closely linking wind farms over hundreds of kilometers in a region such as the Great Plains, basically ensuring constant percent supply and to use smart meters to charge batteries, both stationary and vehicle-borne, when gusts are quite high and stop when they are not.

Worrisome Caveats and Conclusions

One highly respected energy expert, Vaclav Smil (2014), has dismissed any suggestions that renewable energy revolutions can be swift and thorough and that they are "fueled by wishful thinking and a misunderstanding of recent history." Eras such as that of coal in the nineteenth century tend actually to be dominated by the previous periods regarding the prevalence of energy sources. Wood, charcoal, and straw residues from crops globally provided at least 85% of the fuel source for much of the 1800s. Coal was 5% by 1840 but only by 1900 had risen to slightly above 50%. So, the rise of coal took 60 years to reach its dominance. Likewise in the twentieth century with regard to any dominating use, coal was similarly more common even though that is perceived as the era of oil. What he comes up with through intricate historical and quantitative analyses is a claim that a 50–60-year range is required for any energy transition to gain full traction.

At the time of his 2014 writing, only 3.5% of American energy came from the new renewable energy sources of wind and solar, and old hydro had been maximized as a possibility. Even with still ever-rising demand for energy, gas and renewables were barely keeping up. The kinds of efforts that will be required to ensure the equivalent of baseload security from wind and solar will be Herculean.

Another big issue will be that of "stranded assets." Enormous investments have been made in petroleum exploration, oil wells, pipelines, refineries, gas stations, uranium mines and refineries, nuclear power plants, coal mines, and massive coal-fired generating plants. His estimate is that there is at least 20 trillion dollars tied up in them. It is impossible, he claims, for nation-states and corporations to simply walk away from them and their operations since much of the infrastructure, such as in China's case, was expected to last 30 years or more.

He believes, though, that we have been collectively spoiled by energy prices that have been way too low, hiding the true enormous environmental and health costs. He does not offer much in the way of solutions except doing as much as possible to reduce demand, presumably by rising prices and carbon taxes, and then putting out of commission polluting power plants and other sources of greenhouse gases. He is not keen on fast breeder nuclear reactors (ones that reuse already highly radioactive, weapons-grade nuclear waste) or the development of hydrogen as an alternative fuel— fads that both governments have wasted money and subsidies on.

Stan and Priti Gulati Cox (2022) harshly criticize the electric vehicle (EV) and suggest that its widespread acceptance could be *as* ecologically and humanly destructive as the oil industry—causing geopolitical tensions and military conflict. Their focus is on the necessary EV batteries that require enormous quantities of lithium and cobalt, both of which, in the case of the United States, have to be supplied from external sources. The most significant source for cobalt is the Democratic Republic of Congo, and there have been terrible prices to pay over rivalries among warlords over control, with many deaths and much ecological damage. The devastation so far has been in the pursuit of cobalt for cell phones, where a fragment of a gram is needed for each phone, but in the case of EV batteries, each requires pounds. It has been estimated that another 384 cobalt mines would be required to meet the future demand for EV batteries.

Sources for lithium, the other essential ingredient, are found in China and Afghanistan—not exactly reliable suppliers for Americans. South America including Chile and Bolivia have been the alternative sources. Lithium mining requires huge quantities of water for each metric ton produced—500 million gallons leading to heavy losses and pollution in arid regions. Already Bolivia, with its previous left-leaning president Evo Morales, has been subjected to a coup engineered by the United States with the admitted support of Elon Musk, the owner of Tesla Corporation that is dependent upon access to lithium for its car batteries (https://newrepublic.com/article/159848/socialist-win-bolivia-new-era-lithium-extraction).

There has been speculation about exploration and eventual mining for both minerals in the ocean. The South China Sea, a zone of already significant geopolitical tension, is being particularly considered. Altogether that would be environmentally disastrous along with the variety of "green environmental sacrifice zones" that are being created. This is ironic but more tragic as an outcome when you consider the various environmental sacrifice zones that have been created in the extraction of coal, petroleum, and uranium as revealed in Chapters 6, 7, and 9.

Next to critique the possibilities is Andrew Nikiforuk (2023) who portrays traditional environmentalists as having ignored the enormous amount of minerals and metals that would have to be extracted to achieve a 100% WWS electrified world. In a lifetime, the average North American consumes 1.3 million kilograms of minerals, metals, and fuels but basically has no idea how and from where they come and is largely immune to any ethical queries. Mining companies are characterized by Nikiforuk as never being satisfied with existing profits and are lusting after every possible opportunity to extend their operations scraping ocean floors, tundras, rainforests, and eventually considering mining asteroids and the moon. To meet the demand projected for a global revolution through renewable energy would generate a rush on nickel, cobalt, silver, obscure rare earths such as dysprosium and neodymium, and most especially—prodigious amounts of copper. Since about 400 years before the Christian era, humanity has mined 700 million metric tonnes. In order to fully accomplish the new regime, yet another 700 million tonnes would have to be mined by 2040. Copper is already at a shortage and as much as has already been mined would similarly have to be extracted by then. Such mining will be more expensive and consume vast amounts of energy since the geological equivalent of "low hanging fruit" has already been extracted. He also documents some of the severe illnesses and ecological damages that such mining and their tailing pond residues generate.

Parallel to the Coxes with EVs he illustrates some of the costs with regard to specific items in the renewable category. He claims that the average wind turbine requires 500 kilograms of nickel and to refine that amount would burn 100 tonnes of coal. Each crystalline silicone solar panel requires 20 grams of silver paste. To then achieve one gigawatt of electricity from solar panels therefore necessitates 80 metric tonnes of silver in their manufacture. His underscored conclusion is this: "Fundamentally we need to talk about a future of less instead of a future of more." Degrowth economics as described in Chapter 10 are being advocated by Nikiforuk. Note too, this all concurs with the position taken by Philippe Bihouix (2014) as regard to available minerals, and that any mass program to create a non-carbon renewable future that matches our current energy outputs would only exacerbate the very same climate crisis that we are trying to avoid.

What can one conclude after all of these stark criticisms? It is always most certainly wise to pay attention to Vaclav Smil's well-tested realism on any matter concerning energy and here with regard to the long time that it will take for an effective transition to a non-carbon emitting future. Yet in either regime that he cites, coal or oil eras, there was then no perceived climate crisis. There most certainly is one now. So, every effort must be made on policy initiatives to hasten that necessary transition. Green new deals, as described in Chapter 10, come to mind as to possible ways of implementing their complexities. Any transitions, however, will not be without social and cultural upheavals, and for that reason there should be abundant applied, quickly iterative, on-going social science impact research through styles already demonstrated by Gretchen Bakke (2016, 2021) and Laura Watts (2018) to halt, help mitigate, or solve the dislocations.

It is interesting to note that Gretchen Bakke (2021), in spite of Smil's pessimism, indicates her personal surprise at how solar power, minimized in 2010 when she started research, has increased significantly in importance. She gives examples of cloudy Germany and England where, in Germany's case, the sun was providing 7% of the country's electricity by 2018 and 15% when it was hot and dry, and cloudy and wet England is now the world's sixth largest solar power. The "very gray" state of Vermont gains 18–20% from distributed solar. Dry and hot jurisdictions like Arizona, Yemen, Algeria, and so forth could easily produce more than 100% of local needs through the mid-day sun and then store the rest or distribute it further. Much will depend on the nature of public policies on whether or not there will be quick adoptions of renewable power.

As for the Coxes', Nikiforuk's, and Bihouix's revelations, and remembering that they most certainly support an alternative sustainable world, they should be taken very seriously. Implicitly at the very least, they call for very tight international regulations and treaties governing the extraction, processing, of these scarce minerals and intricate public scrutiny in the building of WWS all-electric regimes. Unfortunately, though, given the current hegemony of neoliberal capitalism, the abuses documented will likely continue for a significant while yet. The difficult moral question then could be—would it be better to continue these hugely destructive extractions in the name of a WWS world where electrical personal vehicles dominate or one where we continue "with business as usual" in the oil producing, carbon-emitting regime that we have now? We could hope that any solid green new deal directions might reform public transportation and extend electrical rail service so any felt needs for personal vehicles would become greatly diminished.

Note that many of our commentators led by Gretchen Bakke (2016, 2021) have suggested that the New Grid's restructuring is already underway and is slowly being built from the bottom-up, *locally*, with the goal of integrated and diversified renewable electrical systems. That could be viewed as good news in the form of a significant preadaptation that would be helpful in a likely difficult future. The concluding chapter will elaborate more on this. Under the weight of multiple crises, there is, in my opinion, some possibility for a general collapse of our current global civilization (which in itself may not necessarily be a completely bad thing). For those who might experience such upheavals, this will lead to a set of circumstances that would favor localization and necessitate degrowth toward steady-state economies. So, the more bottom-up locally organized renewable energy systems that have already been put in place the better.

The solutions that this chapter documented before its sober conclusions and others yet to come could serve as a source for assembling local regimes of scaled-down, as ethically designed as possible micro-grids. Such systems of electricity should be designed so that they do not need the kind of technical and scientific expertise that nuclear reactors require nor the enormous capital required for centralized power plants. Gretchen Bakke (2021) in that regard suggests that the new era is turning out to have the character, at least in part, of *reduction* rather than growth in the quest for eliminating fuels and that makes it in harmony, at one level, with that of the ideology of Degrowth. So, there is some measure of hope possible. Another thing that is clear—people will most certainly have to become more patient with the lack of a constant flow of electricity and learn to be more in rhythm to the ebbs and flows associated with nature in gaining from their power sources.

It might have all been much better had our future energy path much earlier followed the expertise offered through the technical mapping provided by our experts and as summarized in the previous parts of this chapter. But that is unlikely now. Let us ethically develop the best that we can in furthering renewable energy on appropriately modest scales in better preparation for young people and our descendants.

Part IV
Conclusions

13 The Big Moral Question

"Buck-Passing"

This problem might aptly be described as one of "intergenerational buck-passing."
  We can illustrate the buck-passing problem in the case of climate change if we relax the assumption that countries can be relied upon to represent the interests of both their present and future citizens. Suppose that is not true. Assume instead that existing national institutions are biased toward the concerns of the current generation: they behave in ways that give excessive weight to those concerns relative to the concerns of future generations. Then, if the benefits of carbon dioxide emissions are felt primarily by the present generation (in the form of cheap energy), whereas the costs are substantially deferred to future generations (in the form of the risk of severe and perhaps catastrophic climate change), climate change may provide an instance of severe intergenerational collective action problem. For one thing, the current generation may "live large" and pass the bill to the future. For another, the problem may be iterated. As each new generation gains the power to decide whether or not to act, it faces the same incentive structure, and so it is motivated primarily by generation-relative concerns, it will continue the overconsumption. Thus, the impacts on those generations further into the future are compounded, and more likely to be catastrophic. If in the long-term there are positive feedback mechanisms, or dangerous nonlinearities in the system (as some scientists suspect), this worry increases.

(Gardiner 2011: 21)

The third part of this book should have revealed that *there can be or at least could have been viable solutions for most of the problems raised here including the Damocles Sword that hangs over it all, us, and our descendants—climate change.*
  Yet, most of these have been theoretical solutions largely only on paper rather than being even implemented, completed on time, or minimally practiced at best. The consequences of our collective inaction accompanied by a hedonistic consumerism could still lead to a tragic outcome for our species and many other life

forms. Dithering, denial, and deceit as well as procrastination and self-interested incapability for long-range planning remain as the foibles and weaknesses of humanity. These failings are denying our descendants and our young people, let alone the impoverished members of the Global South, their fair shares of life-dependent resources and even their very existence, as well as ensuring much suffering to those who may exist in the near future. Those who make the decisions in the capitalist, prosperous, developed North have collectively led many of us (though including many of our very own culpabilities of consumption) along a path into being very poor ancestors.

Morality, Law, and Some Issues Concerning the Young, Future Generations, and Nature

A fundamental concern might be expressed then, you would think, about the injustices being done to the unvoiced—most especially future human generations. Moral philosopher Stephen Gardiner (2011) refers to the situation as the *"perfect moral storm."* In his view, three metaphorical moral storms are coinciding. *The first relates to the anonymity of power whereby those in prosperous developed nations, especially the rich and influential, can shape what is done about the environmental crises and do so to their advantage but to the detriment of those in poor nations and the poor within their own. The second moral storm is that the current generation has similar asymmetrical powers over future generations without obviously any possibility for the reverse. This generation can shape responses to its advantage and cause deep damage to those of the future. The third storm is the lack of political, social, and legal theories and institutions to deal with the first two "storms."*

Governments supposedly protect and solve problems for both present and future generations. Regarding the present, we have theories and institutions that can deal to varying extent with issues such as national security, child welfare, or monetary inflation. Yet we do not adequately have theories and institutional practices to deal with Gardiner's first two moral storms thus the third moral storm. One of the features embedded in current theories blocking that possibility includes the heavy emphasis on the rights of the individual, sometimes even to the extreme of libertarianism. Another example would be the democratic-based expectation of short-term election and policy cycles designed to respond to public opinion but more often cater to the fickle wishes of the electorate—promoting tax breaks, for instance. Gardiner reveals many corruptions in avoiding to act morally on climate change through a clever allegorical comparison from Jane Austen's novel *Sense and Sensibility,* where a wealthy son, aided by the conniving and rationalizations of his equally wealthy wife even after agreeing to his father's dying wish to support his stepmother and half-sisters, still finds self-serving excuses to minimize any supports to the barest possible. The analogies are to rich nations and their responses to the poor developing ones as

The Big Moral Question 209

well as to future generations with regard to doing anything serious about climate change.

He makes a further compelling comparison to our lack of a comprehensive moral theory with a set of ethical responses appropriate to the well-being of future generations. This is done through a brief examination of feminism's struggle against patriarchy. He suggests that a little over a hundred years ago, virtually all of the dominant political and social theories denigrated women's worth—even morally—and in their capacities to make rational decisions such as in the rights of ownership of property and the capacity to vote intelligently. While still struggling against the staggering oppression that patriarchy created, feminism has made significant strides in both theory and practice.

The missing equivalence for the voiceless including our descendants should appear obviously as now underscored and compelling to us but it hasn't. Gardiner (2011) further characterizes this egregious situation in the opening quote of this chapter by labeling it "*intergenerational buck-passing.*" These are important insights since the "perfect moral storm" so far seriously prevents us from acting ethically in the context of the worst set of crises likely to be ever experienced by humanity. We should be, I personally believe, desperately and ethically compelled to work swiftly to remedy this intolerable situation.

Gardiner offers us a tentative ethical theory with eight principles to provide the basis of an on-going discussion of how to respond to the perfect moral storm. They are:

1. Ethical considerations are already at the basis of international climate policy.
2. Scientific uncertainty does not justify inaction.
3. Precaution is theoretically respectable.
4. Past emissions matter.
5. The intergenerational burden should fall predominantly on the developed countries.
6. Specific intergenerational trajectories requires ethical defense. (Here he refers to the various scenarios for carbon-emission reductions by percentages and degree of development by countries over specific time frames.)
7. The right to self-defense is an important, but sharply limited rationale. (E.g., claiming that a sudden change would hurt specific economies such as North American living standards).
8. Individuals bear some responsibility for humanity's failure. (Gardiner 2011: 437)

Regarding future generations, how and where could their issues be best articulated? Claiming their rights in courts of law might be one or even the best method possible, and to use the law to halt projects involving petroleum and coal on the basis of the climate change damages that will be done to such persons, animals, or ecosystems. Yet that is easier said than done when you consider the slow, convoluted, and indecisive court cases that have been completed or are in process. As an example, one of the seemingly devilish arguments already used by defendants such as governments or corporations against allowing future generations to be represented as plaintiffs is the contention that they are not real entities but only theoretical ones because there is no way of knowing exactly who they will be or exactly who regarding membership and in what numbers they will be (Abate 2020). So, what are among the alternatives?

Social movements for one especially those that also voice the grievances and worries of their closest representatives—today's youth. They can at the very least operate as stand-ins for future generations. Consider these rather shocking projections of intergenerational inequities regarding the living young (Thierry et al. 2021) by 28 climate scientists. They compared the probabilities of severe climate events being experienced by children born in 2020 with cohorts born in 1960 (in effect, their grandparents)—first on a global average. These children, alive today and not some future generations, will experience three times as many extreme climate events as the older cohort—twice as many wildfires, two and a half times as many crop failures, and three times as many floods. The authors also looked at regions of the world such as within the Global South that are innocent of the events leading to climate change. They suffered from the most egregious thefts of their resources and sufferings of their peoples (including slavery) from the colonial extractions and appropriations during the Age of Imperialism and continuing today through the extractive forces of neoliberal globalism. Regarding Africa, a comparison regarding the young cohorts is to circumstances of generations during the preindustrial period. The 2020 cohort is expected to experience severe heat waves at a rate fifty times greater than their ancestors and considering that they are dealing with a circumstance of only two centuries. All of these comparisons should be at the very least sobering if not shocking.

Consider also that these are just projections of occurrences and do not include any notions of the *intensity* of these events. Beyond any scientific data, most of us who have been alive at least since the 1950s would attest that the climate events have gotten much more severe and the most obvious conclusion is that they will continue to do so and at severities perhaps yet to have even been experienced before. The three-page, extremely tight article by Thierry and colleagues (2021) does not comment on the sequelae of these events such as the zone extensions of virulent reservoir and vector-borne diseases, the dislocation and suffering of peoples through climate wars, large-scale famines, and mass migrations. The article may even be an underestimate if viewed from perspectives gained from astonishingly negative climate events occurring over the several further years beyond 2021. Its scope cannot cover the intangible and hard-to-express huge losses in the comforts and joys gifted by Nature as well as its capacity to rejuvenate or heal the human spirit that might never be experienced by these innocent victims. For these young people, Nature instead might even come to be seen as some kind of frightening or avenging power. We adults should reflect on these realities every time we encounter a toddler or any young person, remembering that they are not some distant future concerns that we can put out of mind. They are very much of the present and may ultimately be the victims of the greatest ever crimes of humanity.

Greta Thunberg having organized Fridays for Future climate strikes in Sweden and beyond is the best-known and inspiring representative of youth movements. Internationally she speaks passionately, bluntly, and effectively with nothing held back in criticism of politicians, polluters, and petroleum interests at many international climate conventions, various parliaments, on the media, and even the World Economic Forum at Davos. She is well-regarded by the experts and was able to get a significant number of them—George Monbiot, Naomi Klein, Bill McKibben, Michael Mann, Jason Hickel, among many others as authors— for her superbly edited and comprehensive *The Climate Book* (2023) with extensive commentaries by herself.

Another youth organization of significant note is the U.S.-based Sunrise Movement formed during the extremely depressing, climate-denying years of the Trump presidency. Patterning itself on the U.S. civil rights movements, it has avoided privileged, white, middle-class optics since many of its leaders and participants, basically under 30 years of age and including teenagers, are racially and socioeconomically diverse. Tactics used to get public attention have included hunger strikes, protest marches, and occupation of politicians' offices where confrontational interviews with press coverage are held with key politicians who have been negligent of environmental issues. They have successfully helped elect about a score of progressive Democrats and have been instrumental in spreading Green New Deal-styled ideals (see Chapter 10). Sometimes affiliated with the politics of Senator Bernie Sanders and Representative Alexandra Octavio-Cortez, they played a key role in the Democratic Party's factional policy bridge-building after the more conservative candidate, Joe Biden, won the Presidential nomination for the 2020 election. The Sunrise Movement's support delivering over three and a half million, young first-time voters in swing-states provided the largest chunk of the margin of victory for Joe Biden's 2020 win over Trump (*The Guardian*, Nov 16, 2020, https://www.theguardian.com/us-news/2020/nov/16/joe-biden-climate-crisis-ennvironment-energy). Regarding American administrations, Biden's policies have proven to be the most attentive to the climate crisis to date. Yet we cannot be too sanguine because given the nature of American political domination by the petroleum industry—even Biden approved a controversial Arctic drilling program at the time of this writing (*The Guardian*, March 13, 2023, https://www.theguardian.com/us-news/2023/mar/13/alaska-willow-project-approved-oil-gas-biden).

Returning to legal issues, Randall Abate (2020) has written an overview with recommendations for strategies and reforms in his *Climate Change and the Voiceless: Protecting Future Generations and Natural Resources*. An array of North American and international court cases representing present and those children yet-to-be-born, as well as particular non-human species, important landmarks, and particular species are considered. So far for the effort expended, the results could be considered as somewhat sobering.

Some results to date from foreign courts have been more positive than in the United States and Canada (Abate 2020). As examples, the rights of an endangered river dolphin species and a sacred river have been confirmed in India. Certain regions and rivers have been similarly protected in places such as New Zealand. The logic of expanding this sort of protection of wildlife and resources seems eminently fair and responsible, even considering their rights as "persons." As a fiction, that right has been, in so many jurisdictions, extended to corporations that have done so very much of the damage that we are concerned about in this book. Nonetheless, even though more positive than the United States and Canada, those efforts elsewhere still have a very long way to go to make a serious impact.

Still Abate advocates continuing to push forward in the courts, even though plaintiffs may at first suffer setbacks in either not getting convictions or the needed legal protections. It is important to find ways that become legally entrenched allowing citizens to represent, advocate, and act as stewards for those who are currently voiceless–non-human species, landmarks, regions, young people, and future generations. The rationale is that by continuing litigation, legal precedents are gradually being established that will allow courts to eventually consolidate wider laws to protect the voiceless. That is just among the necessary steps required for progress in the evolution of common law as related to the extension of rights and protections.

Abate recommendations are important. Among them he suggests the development of new administrative bodies with legal clout to protect the rights of voiceless entities. For instance, both Israel and Wales have bodies that monitor the potential impacts of legislation on future generations. He advocates shifting from our anthropocentric rights framework, in that resources and wildlife are perceived as worth protecting through their usefulness to humans, to a legal framework that is instead more eco-centric. That would be placing the value on nature for its very own sake rather than how it benefits humanity. As it stands now, it is much easier for a corporation or government to succeed in overturning injunctions that, for instance, try to prevent opening up new petroleum fields by arguing that they are essential for security and prosperity. His recommendation for a rights-of-nature transformation being eco-centric in character would indeed be a significant breakthrough and lead to mechanisms to more directly and effectively deal with climate change. But again, time is very much of the essence.

Geoff Dembicki (2022) in his *The Petroleum Papers: Inside the Far-Right Conspiracy to Cover Up Climate Change* shockingly documents how, beginning in 1959, major oil companies such as Exxon and later through their very own research facilities were aware of the dangerous impact of the carbon emissions generated by their own products. He reveals that they knew this several decades even before it became well-established in mainstream science, by governments, and then filtered to the public. When it became more widely known through general scientific enquiry, instead of concurring and acting ethically, they poured huge amounts of money into lobbying governments to successfully resist regulations on their activities and opened up even more toxic oil fields such as the Alberta tar-sands. They also funded biased research institutes and peoples' front-styled movements that denied climate change or generated time-wasting skepticism that received far too much attention from the corporate media on the basis of "hearing from all sides." Through these fronts, oil interests such as through the infamous Koch brothers, funded successful campaigns that elected Tea Party and libertarian climate change denying politicians to the U.S. Congress. They then blocked American commitments to important international climate change treaties and the petroleum and coal lobbies prevailed, fostering a rise in climate change skepticism among the public. The tactics used in these misinformation campaigns resembled those used by the major tobacco companies and many of the same personnel were hired to conduct the deception. There is no other possible conclusion other than that *these corporations deliberately lied.*

Some have considered these disinformation and distraction tactics, along with the increased production of petroleum and with the opening of even more polluting oil fields, including dangerous and toxic practices such as fracking, as *essentially criminal.* This is argued by the fact that people have already died in considerable numbers because of the intensified nature of climate events— typhoons, hurricanes, large-scale fires, and flooding. Dembicki (2022) gives an extended example of a Philippine climate change activist who became so motivated after losing her entire family in a typhoon-caused flood that carried them all out to sea.

Using a highly sophisticated statistical analysis, Vicedo-Cabrera and colleagues (2021), drawing from a huge data pool from 732 locations in 42 countries from 1991 to 2018, calculated that 37% of 29,936,896 hot season deaths were attributable to recent human-caused climate change with local variances ranging from 20.5% to 76.3%. That is a significant number of over ten million and a number that we can expect to grow considerably as global temperatures continue to rise.

Accordingly, two legal scholars (Arkush and Braman 2023) have published a hard-hitting, 70-page article in the influential *Harvard Environmental Law Journal* titled "Climate Homicide: Prosecuting Big Oil for Climate Death." They liken any potential court cases to those attempting and eventually succeeding in convicting tobacco companies for multiple deaths and public health care costs and the Purdue pharmaceutical company for the lives lost or ruined in the opiate crisis. Regarding American law, they detail the various charges, as examples, of homicide, manslaughter, and negligent manslaughter that the accused could be tried, and the evidence that could be used. In turn, they anticipate the defenses likely to be used and how to respond to them. Like Abate (2020), these authors see potential setbacks at first, but in time precedents would be established regarding, for instance, the types of evidence allowed. Rather than federal courts, they suggest a multitude of actions in state courts, as in the anti-tobacco strategies, allowing for more opportunities to find cracks in the legal system. Eventually, they predict, there would be a tipping point of several successful cases followed by a possible cascade of convictions similar to the tobacco cases. Petroleum and coal companies might then be compelled to suspend operations, be forced out of business, and pay huge fines. The latter in the billions or even hundreds of billions of dollars could be used in necessary mitigation projects to reduce greenhouse gas releases and switch to non-carbon-based economies.

Reinforcing the opinions of both Abate (2020) and Arkush and Braman (2023), the U.S. Supreme Court in late April 2023 rejected an appeal by ExxonMobil, Chevron, and Suncor Energy that would have only allowed litigation by municipalities claiming damage from climate change induced by the burning of petroleum products to be done in federal courts. This appeal had prevented cases from being tried in Rhode Island, Hawaii, Maryland, and Colorado for over six years. Now they will be tried in state courts where the plaintiffs had preferred. Commentators compared it to a dam being broken and led to all sorts of speculation comparable to the Big Tobacco cases (*The Guardian*, April 25, https://amp.theguardian.com/environment/2023/apr/25/experts-hail-decision-us-climate-lawsuits advance?fbclid=IwAR2Yx1CZDFU3pAA5qbb0Vk3P_FH2NPtR5XqalfMk3Icnb8GnKbeCUdBxAhU).

While finishing this chapter in August 2023, added to that success (although not directly involving oil companies as defendants) is an important landmark case involving youth plaintiffs in the First Judicial District Court of Montana. The judge ruled in favor of them.

> Judge Seeley ruled on the scientific evidence presented during the trial. "There is overwhelming scientific consensus that the Earth is warming as a direct result of human GHG (greenhouse gas) emissions, primarily from the burning of fossil fuels," Judge Seeley wrote in her decision. "So long as greenhouse gas emissions stay at high levels, the Earth will keep getting warmer and the climate impacts we are already seeing – wildfires, drought, sea-level rise, extreme heat – will keep getting worse."
>
> In her ruling, Judge Seeley emphasized that the young people will disproportionately carry the weight of the climate crisis. "Children born in 2020 will experience a two to sevenfold increase in extreme events, particularly heatwaves, compared with people born in 1960," wrote the judge. (Snyder 2023, https://www.counterpunch.org/2023/08/25/the-climate-crisis-goes-to-court-youth-prevails/)

Any and all attempts to change legal systems toward guaranteed protections of future generations, Nature, and specific non-human sentient beings are certainly welcome. Yet more immediate and direct lawmaking by legislatures and parliaments rather than just the courts is urgently required. Laws are needed that ban the extraction and sale of fossil fuels along with the appropriate regulatory agencies to enforce them, speeding up the accomplishment of no more carbon emissions. Assisted by legal strategies, we need a concurrent mass mobilization of the public and one as large as never seen before. Otherwise, politicians will continue their procrastination and catering to corporate interests. They need to directly legislate laws that will compel zero-carbon economies as rapidly as possible—thus establishing green new deals if not genuine degrowth and in international binding contexts.

Movements and Mass Action—Manifesting "Climate X"

From Chapter 10 recall that Mann and Wainwright (2018) label their preferred yet unconsolidated means to solve climate crises as "Climate X"–mass affiliated and determined social movements emerging from the global grassroots. By sheer numbers and coordinated actions, they would force governments to respond or they would directly take over where nation-states had failed. What currently is the state of that possibility?

Since the early 19th century, social movements of a largely secular character have taken the lead in bringing about social change in the context of the disruptions of industrialism and globalization, and cascading calls for justice. It would be hard to note a single progressive change that has occurred in the modern period that has not been brought about by the actions of dedicated social movements (Ervin 2015). We can take inspiration from the examples documented in this book. La Via Campesina in agriculture and The Coalition for a Clean Green Saskatchewan and The Committee for Future Generations in resistance to nuclear energy representing the large, intermediate, and small scales respectively. Consider once again the remarkable example of Greta Thunberg, who as one single teenager, built a mass movement regarding climate change on the basis of her lonely Friday afternoon school strikes in front of the Swedish Parliament.

Seth Klein (2020) has suggested, given the scope and intensity of the crises, it should be metaphorically "all hands, on deck now" We are in an extended and complicated crisis with many fronts and local and general issues happening all at once as well as having to deal with the negative tipping points that are rapidly accumulating. Massive social movements need to provide a counterforce to the hugely destructive influence of the corporate sectors—most especially those extracting fossil fuels. It is also imperative that we do not entirely leave the burden for all of this on the young. Such an international alliance of movements should be, while on a scale never before seen, intersectoral in that it includes most other justice-seeking groups such as women, youth, unions, the impoverished, indigenous, the deeply concerned and alarmed, and especially those from the Global South who have carried the brunt of climate change so far. It should reach such a tipping point of demonstrable public will that policymakers have no other choice but to act decisively, quickly, and comprehensively—putting in place the radical transformations that are now required to create a post-carbon, environmentally responsive civilization. Most likely civil disobedience, even at its maximum levels, might eventually be required. Taking such a radical position, Andreas Malm (2021) author of *How to Blow Up a Pipeline* underscores that point. While he does not really give instructions to actually destroy pipelines, his argument is that the ruling classes and politicians are so enmeshed with the status quo represented by the immense power of the petroleum companies that property damage beyond just non-violent civil disobedience such as occupations may be necessary. He *does not*, it must be stressed, advocate the taking of lives.

Relatedly though, the late French philosopher-anthropologist Bruno Latour along with his Danish acolyte, Nikolaj Schultz (Latour and Schultz 2022), has some important things to say about difficulties of such movements and the shortcomings of what they label "political ecology." Their version of the subject differs in some regard to that presented in this book you are now reading. The latter, the academically-based version that I have presented here, while most certainly appreciating and advocating political action, puts more emphasis on what is critically known through research and documenting what could be solutions. Alternatively, Latour and Schultz (2022) highlight the *political* or *action* parts to bring about change through movements, laws, and policies, most especially through the legally mandated programs of representative, liberal, or social democratic governments. They appear to be referring to what we might consider normal political structures operating through elected parliamentary systems that have political parties representing the different sectors and interests within society. Parties operate in competition and sometimes cooperation with some emerging as winners achieving the function of governance through the achievement of legitimate electoral power. In such circumstances, integrated policies mitigating the environmental crises such as a green new deal or one of degrowth and redistribution as discussed in Chapter 10 might be put into full practice. That would be for the sake of solving the multiple environmental perils looming over us.

Yet they lament how little change these forms of "political ecology" have achieved. In the context of Europe, which they would know best, green parties that should represent the vanguard of what Latour and Schultz consider to be "political ecology" have achieved almost no electoral success of any significance and have been far from achieving governing power. They have sometimes been part of governing coalitions in countries with proportional representation systems but in minor non-binding somewhat token roles. They have always tended to be marginal in the number of seats and votes gained. Latour and Schultz

(2022) exemplify this lack of success with a recent French national election where the Greens failed to even regain their financial pre-election deposits since they had less than 5% of the vote.

The urgent task for Latour and Schultz's version of political ecology is to develop a widespread collective identity and shared belief system equivalent to a mass *class consciousness* among those affected or concerned by the crises. Then they would enter the political arena with an explicit agenda built on a *class struggle* knowing exactly who both their enemies and allies were and organize disciplined campaigns directly appealing to their logical class constituents. Historically, they point to the rises and successes of the bourgeoisie during the spread of modernity through business-friendly, classically economically-liberal oriented political parties and the ascendancy of capitalism, and then the working classes through labor and social democratic and socialist parties in responses to the dislocations of the Industrial Revolution. It is the political parties in their many varieties that are pushed into formation and action by social movements rising with such classes where a meaningful consciousness has been formed.

So, who would constitute such a yet-to-be-formed-or-completed class? Latour and Schultz suggest the young—surely, they would have the most direct stake or "skin in the game" as the colloquial expression goes. They mention only a few other potential categories. One is that of indigenous people who they suggest represent about one quarter of the world's population—assuming that they include the vast numbers of say peasant populations found in Asian countries such as China, India, and Indonesia. In their cases, typhoons, heat waves, crop failures, and flooding could make them candidates for the class consciousness Latour and Schultz (2022) suggest. Their political clout would depend upon their level of consciousness about the cause of these climatic events and the degree that the political cultures of their home countries allow for them to have influence. Another potential constituency mentioned is that of the wide "swath" globally of intellectuals and presumably that could include a large portion of natural scientists who surely must be fed up by the lack of action considering their urgent warnings and recommendations by policymakers. They also, without elaborating, suggest that we should not forget the "religious" as a potential source of supporters—possibly on the basis of deep moral concern, I might speculate.

As part of the problem for forming this class and its consciousness, they refer to the twentieth century as one where values of prosperity, growth, liberty, along with relatively few notions of limits dominated. Among the prevalent political classes of liberals, social democrats, socialists, and conservativesthen, there is still little understanding or identification with the concerns of those in this yet-to-be-fully-formed "political ecology class." The former categories are the ones that largely determine the agendas of political parties forming governments. There is, of course, even among most of these categories, a moderate to vague understanding that global climate change and other environmental degradations are something to be seriously concerned about. Latour and Schultz (2022) provide a clever analogy, but with perhaps not a particularly sensitive terminology—collectively regarding our climate change present and future, we are as helpless as "primitives" trying to understand and deal with modernity.

This potentially emerging political class with a distinct consciousness and building an agenda "from below" should question the very foundation of modernity itself with its overriding emphasis on "development." Counter to this, the new *emphasis for a major transformation should be on "envelopment."* While they do not elaborate on the meaning of envelopment, I take their assumptions to mean an eco-centric world with steady-state economies and with satisfaction accomplished through reforms as roughly commensurate with those of Part III of the book you are currently reading.

They point out that the nation-state was conceived in order to allow the dominant classes to continue to consolidate their control of power. That has to be challenged toward a new view that focuses on **"habitability for all."** For this in expanding an emergent class and its consciousness they suggest a process of building a reinforcing a counter-culture involving the humanities, art, and literature. Eventually through dogged persistence, as I read them, Latour and Schultz (2022) would hope to see a new "hegemony" whereby all of the needed transformations can eventually occur because of a now effective "political ecology" of action and a newly dominating global ideology. A fascinating set of perspectives for sure, but once again, time is of the essence since for instance they point out, drawing from E. P. Thompson's (1991) classic *The Making of the English Working Class* that it took about 100 years to fully shape the working class in its full consciousness.

More within a framework of familiarity for myself and most readers would likely be the global North of developed countries. So far, the majority of the opinion polls suggest that people *do* recognize climate change and ecological degradation as issues—but ones to be put further down on lists to other issues such as employment and cost-of-living concerns. There is a subtle but I think possible undercurrent to perhaps most adult thinking—I know, alas, that even I have experienced it from time-to-time. That thought is that basically while terrible though the future might be, it will not badly affect me because "I will be dead and gone by then." While not having that as an excuse for non-engagement, it is quite possible that even a majority of the young are not really engaged either. This could be for a multitude of reasons including heavy debts and worries for their personal rather than collective futures.

Some encouraging signs, though, have been emerging in the United States of a wider alliance of similar class interests with the formation of an organization called *The Third Act* (https://thirdact.org) in 2021. It was founded by prominent climate activist Bill McKibben focusing on the energy, resources, experience, and available time of seniors—those over 60 who are, as it were in the "third act" of their life after first growing up, then becoming young adults with careers and raising children. Besides racial justice, the most intense focus is on the climate crisis. The organization has local chapters and has been organizing demonstrations against banks with divestment from oil enterprises as a main objective.

Another one founded by Bill McKibben is 350.org (https://350.org/) that is international in scope and membership and has been committed to keeping carbon dioxide in the atmosphere to 350 parts per million and shifting to 100% renewable energy. Extinction Rebellion or XR (https://rebellion.global/) started in the United Kingdom but since goning international has relied on massive demonstrations and civil disobedience but is now moderating its tactics somewhat. It holds that we should reach net zero carbon emissions by 2025. Greenpeace (https://www.greanpeace .org/global/) is international and represented in more than 50 countries. While having goals similar to the above movements it also is notable for notions of social justice in advocating compensation for those places primarily in the Global South that have been most affected by climate change. Highly significant is the Indigenous Environmental Network (https://www.ienearth.org/) because First Nations People serve on the very frontline in protecting lands, water, and communities from environmental damage while being at the same time major victims of corporate and nation-state extraction damages. These examples mention only a few and there are thousands more of varying sizes and scopes.

Geographer Anthony Leiserowitz (2020) at the Yale Program on Climate Change Communication (https://climatecommunication.yale.edu) along with colleagues there and at George Mason University (Leiserowitz et al. 2021) have been tracing American public opinion about the threats of climate as well as in many other counties. A significant part of their objective is to generate strategies for communicating with the public to promote a general *public will* in sufficient scope and depth that will compel governments to take significant action to mitigate the environmental crisis.

From their detailed surveys, they identify six American "publics" revealing opinions about climate change (Leiserowitz 2020): the first is the *Alarmed* represented by 21%, who are certain that it is human-caused and urgent as a threat, strongly support the notion that actions should be taken but most don't know what they would be. Category two is the *Concerned* at 30%, who believe that human-caused climate change is happening and is serious but still consider it a distant problem and something that will more affect plants, insects, polar bears, and people several generations away but not predominantly in the United States. While supporting mitigating policies they don't see them as being urgent. The *Cautious* at 21% are not sure if it is happening and human-caused or if it is overblown as a concern so are characterized as not having made up their minds one way or the other. Coming next are the *Disengaged* at 7% knowing little about the issue and not hearing anything about it from family or friends. Then follow the *Doubtful* at 12%, who think climate change may be happening but it is part of a natural cycle. The final and most hostile are the *Dismissive* at 9%, who not only do not think that climate change is happening but even endorse conspiracy theories about those who believe in it—as a get-rich hoax for people such as former Vice-President Al Gore. While the smallest portion, this lamentable category is very vocal with powerful allies in the U.S. Congress, state legislatures, and sometimes even the White House.

For a successful campaign of communication and action (Leiserowitz 2020) suggests first focusing on the Alarmed with its approximately 53,000,000 citizens, of which 3.7 million are already part of campaigns trying to sway politicians into action. There are another 14.8 million who say that they "certainly would join a campaign"—and another 19.6 million who said they "probably would join." As he points out, the total represents an "enormous potential social movement." More of them need to be recruited and put into action.

All the current movements organizations, political parties, and media outlets dealing with the issue, large and small, Leiserowitz suggests, should now be organized into an *advocacy coalition* with significant clout to overcome the interference of opponents of climate change action. As it stands now, these latter opponents, the Dismissive, through the efforts and funding of fossil fuel billionaires and their better-organized coalitions of research institutes and front organizations, have been able to prevent any meaningful action (Dembicki 2022; Oreskes and Conway 2011).

After the building of a much bigger coalition of movements, green activists need to forge a massive attitude of "silent permission" among the middle class American public, who either recognize climate change or do not deny it. Unlike the Alarmed, such passive public are not going to lobby or participate in demonstrations but they could tacitly support more indirect but insistent action through the electoral process by supporting climate change candidates over those that deny the necessity for action. They need to be convinced of the desirability of such election results and resulting legislation to drastically curb carbon emissions.

Regarding the mainstream American public, Leiserowitz (2020) comes up with a key recommendation for communication. Since climate change alarmism is associated with scientists, liberal politicians, and environmentalists, and while having respect for science generally, Americans, however, do not as a whole identify as liberals or environmentalists. It is important to find spokespersons who represent a much wider range of experience—clergymen, military leaders, businessmen, grandparents, and from all sectors of society. The need is to appeal to that majority that will provide the "silent permission." Finally, all organizations in the hopefully united advocacy coalition should pay much attention to the deep and pervasive training of engaged citizens drawn from the Alarmed public to expand the movement. In his view, few organizations do enough of that now, spending much more attention on legal challenges and lobbying of elected officials, which of course should still continue, but the expansion of the engaged citizenry is paramount to get the movement expansion needed to create a solid public will for climate change action is required as soon as possible.

Beyond Leiserowitz's (2020) suggestions, a much larger equivalent to Latour and Schultz's (2022) class consciousness would still likely arise eventually more on its own once the damages done by climate change become even more widespread, deeper, and visibly apparent.

Conclusion

Yet with all the legal cases underway and the admirable efforts of climate change movements, they may be too late for the kinds of changes and timing that have been recognized as urgently needed by the IPCC. It is perhaps time to consider what might happen next and seriously prepare for it.

14 What Might Happen Next ?

Carbon-fueled capitalism is a zombie system, voracious but sterile. The aggressive human monoculture has proven astoundingly virulent but also toxic, cannibalistic, and self-destructive. It is unsustainable, both in itself and as a response to catastrophic climate change. Thankfully, carbon-fueled capitalism is not the only way humans can organize their lives together. Again and again throughout our history, we have shown ourselves to be capable of shedding maladaptive systems of meaning and economic distribution, developing resilient social technologies in response to precarity and threat, and to transforming obsolete social practices into novel forms of life.

Roy Scranton, Learning to Die in the Anthropocene: Reflections on the End of a Civilization. 2015. Page 23.

A *New* Green New Deal?

This chapter stresses *sufficiency* or just enough in perceptions of well-being; eventual substantive *degrowth* and the establishment of relative *steady-state economies* with very much lower impacts; and a turn away from globalization to *localization*. These principles, however, will not be accepted by anywhere near the majority needed—especially in the Global North. Yet within a hundred years or so, circumstances may well compel them into existence. For the sake of our descendants and our roles as better ancestors than we have been so far, it would be best if many more of us would work toward persuasion and policy solutions that address them.

Where to start—most likely through modified green new deals that would provide carefully graduated transitional steps toward degrowth. The possibilities for green new deals are already accepted to a much greater extent by the public and a wide range of progressive political parties. To get the job done, there needs to be coordination and consensus between governments and social movements. Green new deals, though, currently suggest the necessity for high production to massively transform to non-carbon economies. As we have seen, however, this cannot be completed without an immense throughput of materials and energy along with unintended human rights abuses that would exacerbate the very same crises we are trying to avoid. For these reasons, the substitution of internal combustion vehicles by massive numbers of electric cars is also probably out of the

What Might Happen Next? 223

question—expansions of public transportation would obviously be more effective. Legislated green new deals, though, could provide the necessary authority, funding, and momentum along with highly motivated regional movements to help bring about the infrastructures for modest local systems of diversified renewable energy, production, and consumption. These then could serve as the basis for preadaptation when the climate crisis is fully realized.

Stan Cox (2020) in the *New Green Deal and Beyond* and in a *Counterpunch* article (https://www.counterpunch.org/2019/01/17/that-green-growth-at-the-heart-of-the-green-new-deal-its-malignant/) has explored similar ideas because any degrowth agenda requires government action. The original American New Deal—the source of inspiration for green new deals—was implemented at a time when there was a perceived need for huge expansions in the economy. The very opposite is true now. Cox suggests a *New* Green New Deal. One idea is "cap and adapt" in place of "cap and trade," favored by green growth advocates and some green new deal formulations. The latter establishes periodic reduced quotas on allowable carbon releases but rewards businesses that are able to decrease below their quotas and then sell the remainder of their allowances to less compliant or efficient enterprises. Recognizing that reductions cannot be accomplished all at once and so graduated cutbacks are called for, conversely though Cox insists that *no trading be allowed* and that all quotas be firmly held to with punishingly heavy fines and other penalties applied to the offenders to the point of putting them out of business. In his "cap and adapt" approach, quite significantly, he suggests that the policy be extended beyond petroleum and carbon emissions to *all* major extractive industries in order to shrink the economy.

To merge green new ideals with degrowth formulations then still requires much research and debate, small-scale experiments, public education about its necessity, and the generation of millions of implementation details. There is a need to work out very careful mitigation or compensation solutions for those whose lives will be disrupted. Ultimately, though, only degrowth can be the most effective ways to give future generations a better chance of not only survival but even of well-being based on low impacts, relative equity, and "just enough" or sufficiency in material needs. We could speculate that as degrowth policies and information about their necessity are gradually introduced through government channels along with the visibly obvious ongoing climatic shifts that the public would become more inclined to accept them.

Regarding the future, I take a view similar to Bill McKibben's reflections on possible outcomes in *eaarth: Making a Life on a Tough New Planet* (2010). He suggests that global climate change is well underway and humans will now have to continually adjust to new and sometimes extreme circumstances. The position is similar to predications by Wes Jackson and Robert Jenson's (2022) *An Inconvenient Apocalypse: Environmental Collapse, Climate Crisis and the Fate of Humanity*. In *Learning to Die in the Anthropocene: Reflections on the End of a Civilization*, Roy Scranton (2015) uses an extended meditation about the dying of this global civilization as an analogy to a samurai or soldier meditating on his possible death as a practical and determined preparation for what inevitably has to be done next.

These suppositions are plausible, because in 2018 the International Panel on Climate Change (IPCC) argued that there weres only 12 years left to reduce carbon emissions to prevent global temperature rises above 1.5° Centigrade—considered *the* dangerous tipping point (https://www.ipcc.ch/sr15/chapter/spm/). What has been accomplished by the time this book has reached your hands? Plus, consider that in 2022 the IPCC suggested that continuing at the current rates of carbon release would mean that the atmosphere will be constituted with 550 parts per million of carbon dioxide, compared to 350 ppm that was originally considered the safe point for keeping temperature rises to 1.5°. Although the measurement fluctuates, we were at 424 ppm in April 2023 (https://www.co2.earth/daily-co2). Note that humanity's preindustrial experience was always below 300 ppm and that the gain occurred in only about 200 years out of about 300,000. So, passing a temperature rise of 2° or 3° by 2100 is not hard to imagine especially when considering that the burning of coal has not been significantly reduced and sometimes has actually been increasing (https://www.cnn.com/2022/12/16/world/coal-use-record-high-climate-intl/index.html). Relying on expert opinion, Greta Thunberg (2023) suggests that the temperature rise by then could be 3.2° Centigrade. These temperature rises will bring about transformations that could eventually create a virtually "new" but less abundant planet-regime with conditions different from the 11,000-year period that led to the development of our high-consumption global civilization.

I also think that this future might then portend the disappearance of most of the infrastructures that support globalism—much of its mass-scale industrialism, excessively long globalized value chains with huge international trades of commodities, and the viability of neoliberal capitalism that holds continuous growth as mandatory. The real possibility of international financial collapses seems previewed by crises such as the global bank and investment firm breakdowns of 2008–2009. The existence of under-regulated financial instruments—especially the derivatives markets that preposterously can in total make claims on assets that are now ten times the world's annual gross domestic product—and other burstable bubbles continually threaten international financing. Pressures upon national governments to bring about stability through taxpayer-funded bailouts may accumulate to a no-return breaking point. Climatically and environmentally, the storms, floods, massive fires, droughts, crop failures, and heatwaves that have been coming our way are becoming much more than simply too expensive to deal with. Again, they are stretching the debt capacities of governments and financial institutions—most especially the insurance sector. The carbon-based energy sources that have fueled almost all of the spectacular surges of growth, lasting now for two and half centuries, will have reached a point where

What Might Happen Next? 225

they will be too expensive or otherwise come to an end by the cost and crises brought about by trying to manage climate disasters.

Events such as these may well generate for those several generations from now *realities that can only most of the time be effectively adapted within completely new frameworks*. What follows next is a set of suggestions for dealing with what could be considered close to worst-case scenarios (extinction would obviously be the very worst). At the same time, it is my opinion that if followed they would have the advantage of preventing the very same near worst circumstances.

How Could People Respond to the Scenario of Global Warming and the Collapse of a Global Civilization?

So, here is an exploration of a few principles that might guide the adaptations of a necessarily much more resilient human species on a much diminished, lower case "eaarth" as Bill McKibben (2010) renames it—implying a lesser capacity to benignly support humans and other life forms. It likely could no longer maintain the markedly unequal prosperity of the developed Northern parts of the globe but was denied to so many others. The pumped-up turbo-capitalism (all engines-no brakes) global civilization that had been built around the international trade of extracted resources and manufactured products from distant sources, once again, likely would have eventually collapsed or become significantly diminished in its capacities.

What about the nature of socio-political units? A case could be made that *localization* rather than globalization (see De Young and Princen 2012; Norberg-Hodge 2019; Rees 2014) might be necessary because of the restrictions of transport, trade, and the collapse of long commodity chains. Whatever is equivalent to many states might be limited in size—say Denmark-sized. They would be strongly influenced by local circumstances such as natural land barriers, boundaries, and historical contingencies such as the degree of ethnic and national solidarity. City states may emerge out of larger entities and draw from smaller hinterlands of agriculture and sources of other raw materials. With reduced populations and scaled-down industries, they would be less prosperous but more resilient and that would be the most important factor when considering the scale of crises that preceded their establishment.

Localization is related to the principle of *subsidiarity* wherein problems are better solved or prevented at the places where they occur. Local planning could more effectively generate solutions to the primary needs of providing materials, food, water, and energy. Bioregions and watersheds come to mind because those within would be motivated to cooperate in territories that they directly comprehend for managing resources. Related to that could be smaller scales of societal interconnection allowing self-organization to effectively emerge in the creation or *re*creation of more nimble institutions and practices that foster resilience and adaptation. William Rees (2014) points out that human minds are incapable of fully comprehending the massive scales and complexities of our current crises and how to respond. With localized bioregions and more discernible carrying capacities and limits in resource use, they would be in much better positions to respond more effectively.

Ideas about the advantages of small-scale political units were earlier promoted by the Austrian economist Leopold Kohr's (2001—originally published in 1957) *The Breakdown of Nations*. Now just because something is a good idea doesn't mean that it will come to pass especially in a uniform manner. However, in today's world, there are separatist tendencies at work in many political jurisdictions even with larger federations such as the United States, Canada, and Russia. How the tendency would play itself out in this scenario is anybody's guess. The dangers from invasion by imperially motivated states might still remain especially over access to resources and more benign eco-zones. Yet the first compelling need may well be to get one's local adaptations established in order to survive in what could be an extended and deep crisis. These scenarios do not mean that whatever political entities emerge will completely isolate themselves and not cooperate or trade, but the frameworks will be radically different than today's.

Most hesitate to make predictions about human population sizes but local carrying capacities and other geopolitical and above all Nature's responses would be much more determining factors than now. Many nation-states in the developed world frequently exist far beyond their carrying capacities because of large per capita ecological "footprints" where they benefit from resources from other regions. Wackernagel and Rees (1996) estimated that the Netherlands, a small, densely populated, prosperous manufacturing and trading-dependent nation, had an ecological footprint that was fifteen times its national territory! That would be impossible in this new situation. Very much more emphasis on local sufficiency would be imperative.

And attention to *resiliency* would be mandatory. That would likely involve a change to localized, low-impact contexts of flexible ingenuity with small-scale and appropriate-to-newer-function technologies from lessons already learned and much still to be learned or relearned. Resilience means the capacity to keep returning to previous roughly steady-state conditions or "bouncing back" after repeated setbacks. As an example, Cuba in the 1990s, after its food and energy crises made extensive use of animal traction and natural fertilizers while using extensive scientific knowledge (Rosset and Benjamin 1994). Farmers applied animal and green manure fertilizers, crop rotations, polycultural systems, the use of insect and plant equivalents for pesticides, along with a newly trained labor force with a focus on cooperatives and small-scale market exchanges emphasizing mixed crops and livestock rather than sugar that had served as the dominating exported monocrop. Near complete self-sufficiency was accomplished, and, now much more resilient, Cubans are no longer as subject to the vagaries of world markets.

This segues into *transitions* and *cooperatives*. The experience in Cuba represents what was an incomplete transition but one that could be revived. The relevant plans were developed in anticipation of an energy crisis and more was devised during the process itself. Similarly, *transition towns* begun in the United Kingdom provide another experimental and information-collecting movement—through community gardens, permaculture, and alternative green energy all focused on localization and sustainability (Hopkins 2014). The study of *cooperatives* would uncover yet another set of rich principles. It would be wise to document and analyze current and past attempts at sustainable living such as small-scale intentional communities, experiments in permaculture, and renewable energy. The internet is full of sources about alternative solutions— the Schumacher Institute is one excellent general example—https://www.schumacherinstitute.org.uk or; another is the Post Carbon Institute https://www.postcarbon.org and https://www.resilience.org; Global Ecovillage Network https://ecovillage.org/; the Rodale Institute for Agriculture—https://rodaleinstitute.org the; and Solutions Project with regard to energy—https://thesolutionsproject.org.

Bill McKibben (2010) enthusiastically supports the retention of the internet as a vital tool of communication and adaptation in his version of scaled-down, no-growth, frugal, circular, "hunkered-down," and largely localized economies. The biggest problem with the internet would be the huge amount of energy required to keep it operating, especially in its vital information storage capacity. Bihouix (2014) informs us that about 10% of global electricity use is now devoted to information technology. How it all could be redesigned in a less globalized world with much less energy remains to be solved, but its retention could be useful because it already contains valuable instructions on such domains as horticulture and repairs of small-scale technologies. It would be valuable for the sharing of any future solutions in a context of widely dispersed localized societies.

Conviviality in a Low-Tech Society should be a principle governing adaptation. Ivan Illich (1973) in *Tools for Conviviality* referred to tools as implements as with hammers and saws but also to institutions such as education or medicine when used to achieve some explicit rational purpose. There are many layers to this important concept—only a few can be dealt with here—and I will focus on some dealing with implements or complexes of them rather than institutions. The crux of the problem leading to the climate change crisis is the massive throughputs of the Earth's material resources based on the combination of our powerful technologies fueled by huge amounts of carbon-based energy. Any future will have to be low-tech and use far fewer quantities of natural resources because shortages would be evident and could not be compensated by long-distance trading. More easily constructed, repairable, recyclable, and simpler technologies within the knowledge and skill set of the average person while serving the larger community and the local environment will be preferred. Community and regional energy sources would be convivial if they consisted of solar panels, wind turbines, and run-of-river sources. Nuclear reactors would definitely not be convivial. Bicycles, cargo bikes, and composting toilets would be convivial. Automobiles would not. Commodity chains should be much shorter, and smaller-scale factories should be built close to localized populations (see Bihouix 2014; Vetter 2018).

For those who value democracy, equity, and egalitarianism, *emergent bottom-up, cooperative behavior* would be the source of preferred action—as opposed to top-down authoritarian compulsion. During the Great Depression, the Canadian Prairies were hard hit with its dust bowl and an impoverished farming economy. This was the region where rural and small-town municipalities pioneered community-based medicine that later consolidated into Canada's national health care system (https://canadaehx.com/2020/06/27/the-road-to-medicare-in-canada/). During that time just about any retail sector and agricultural marketing was organized through cooperatives. Coops are oriented toward building, preserving, and managing resources in local regions and much less tied to the whims of banks and exclusively market and profit logics. Coops for organizing local energy systems using wind, solar, and micro-hydro sources are logical. Public and credit-union styled banking are options that are more suitable to a post-neoliberal, post-carbon, post-corporate dominated world in that financing is focused on the needs of the regional communities enhancing stewardship of human and natural resources.

Yet another principle is *reclaiming the commons* (see Bollier 2014; Bollier and Helfrich 2012). Previously for peoples around the world and as vital parts of their cultures, commons meant *shared* territory, resources, and bottom-up rules, regulations, and management networks for sustainability. Scholars such as Bonnie McCay and James Acheson (1987), and Elinor Ostrom (1990) have demonstrated the effectiveness of local commons for organizing irrigation systems, underground aquifer access, fishing grounds, pastures, forests, and shifting agricultural lands. New domains for commons have already been opening up—much that is available through on-line social networking, shared software, and on-line encyclopedias. In many places, such as with Indian forests, Mayan *ejidos*, and Brazilian unoccupied land estates, local peoples have been *reclaiming the commons* and reshaping cooperative relations for shared management of resources. In some situations, older cultural elements are rejuvenated and in others created anew. Ultimately reclaiming the commons in local contexts ensures that people are in social relations that encourage them for mutual well-being to locally economize for themselves and future generations. Using commons procedures for food production and the collection of other needed materials such as fiber and minerals are logical examples. One clear advantage is that the use of the commons as a method of managing and distributing resources promotes equity and the reinforcement of cooperation during very difficult times.

Michel Bauwens et al. (2019) in *Peer to Peer: A Commons Manifesto* outline an ambitious plan in which "commons-based peer production" could constitute a major historical transformation in political economy—possibly eventually replacing capitalism as *the* major mode of production. Its underlying motives would be socially based and focused on widespread sharing, and be mutually "generative" rather than capitalistic, extractive, class-biased, private property-oriented, profit, and accumulation based. It would benefit from the internet and expansions of open-sourced plans, procedures, and designs for innovations. Given how prolonged the climate crisis might be, that could be an alternative for international cooperation since trading supply chains would be limited.

Neoliberalism and neo-classical economics have miserably failed us and have instead led us to our multi-daggered global crises. Growth would be quite unsustainable in these new circumstances. ***Degrowth with circular and steady-state economies would be forced because the widespread crisis contingencies would compel their shrinking.*** More or less stabilized economies were abundant in the past so they cannot be brushed off as being impossible to put into effect under these conditions. Our current economies, dependent on constant growth, based on the banks' fabrication of credit that fuel capitalist enterprises and ever-constantly growing consumer purchases and debt, and the casino capitalism of the financial sector, insanely necessitate the infinite enclosure and commodification of as much of the world's resources as possible.

While new economic models are needed, this would not mean the complete end of entrepreneurship in business or of all private property. They, however, along with the notions of growth and money, could no longer be fetishized and would be subject to tight regulation dealing with real scarcities rather than artificially induced ones that are now part of the consumer society. Markets would serve utilitarian functions of allocation and distribution rather than for unfettered gains in personal wealth.

How might people contend with scarcities regarding materials for building homes, infrastructures, and the likely drastic reduction in energy sources? Of course, the latter will depend on the local realities of what is available. Bottom-up voluntary simplicity and frugality managed by such institutions as commons and cooperatives generated by a culture of consent also bolstered by recycling, repairing, and reusing should suffice and be preferable in most cases. *Rationing* with quantities clearly specified seems like a logical tool for that enforced by a local commons type of resource management or by whatever local governance form takes shape.

Of course, all of these above suggestions remain purely speculative, because nobody can *ever* make guarantees about something as elusive as humanity's collective future. Please consider that all of the above, for the moment at any rate, just represents one person's (albeit educated) opinions—i.e., mine. By all means debate them—as to their reality and practicality.

Conclusions and Final Words

Nonetheless, there is no doubt that the future will be more than just challenging. The possibilities may lie between small, regionally linked, bottom-up, democratic, cooperative, political entities but also contrasting large "Climate Mao"-like authoritarian scenarios and many variations in between. There would have to be a lot of thought and patient experimentation put into play for attempts at softer landings. One would have to further use their imagination to speculate on the varieties of social and environmental realities facing our descendants. And the multitude of yet unknown climate change consequences let alone the standard human foibles, including procrastination, willful ignorance, vested interest, ideological distractions, stubbornness, and intractability, that may stifle many constructive adaptations. It would be hard to avoid the conclusion that territorial security and all that implies could amplify the perils. What about climate refugees—from where and in what numbers?

I have presented these more somber speculations as different than the otherwise understandable attempts, written in order to get people on side, that show the future could be bright with green, non-carbon solutions. Time is running out though. So, the above scenarios are plausible although obviously by no means certain. Indisputable though is that since we already know that climate events are going to be difficult for current young people and future generations, we are morally compelled to reduce any further calamities. That means following through with our maximum efforts with all the strategies mentioned in the previous chapter—massively integrated global social movements, aggressive electoral action, boycotts, civil disobedience, relentless legal challenges, new green deal legislation, and whatever else can be mustered.

Consider for every fraction of a degree of temperature increase avoided, that probably means the savings of millions (maybe hundreds of millions) of lives. In the Australian bushfires of 2019–2020 attributed to global warming, it was estimated that 430,000,000 mammals, birds, reptiles, and frogs died (*BBC News* Jan 4, 2020, https://www.bbc.com/news/50986293). So, I am not just talking about humans, but specifically regarding them, I ask you to imagine their possible futures every time you meet an infant or toddler.

In this current age of globalism, we might still hope that there will finally be a united international effort through what Mann and Wainwright (2018) called "Climate Leviathan" acting with its collective sovereign power to stop carbon emissions. But regarding the rest of this uncertain future and any further solutions, they cannot be generated through a grand plan to be applied uniformly and globally. That has been the problem with modernity and globalization projects that were taken over by growth-obsessed economies. Solutions would be better if they are eventually uniquely assembled on a bioregional, localized basis—situation by situation. The parts need to *first meet local realities and traditions* and *there is enough knowledge already out there to assemble workable solutions.*

Such necessary experiments in living should include traditional local ecological wisdom and, above all, counter the overwhelming dominances of the ideologies of extractivism, productivism, consumerism, growth, and the domination of nature, that if continued, will prove our downfall as a species.

There is a hope among many of us that, somewhere, sometime, there can exist a kind of society where humans live in a harmonious, sufficient, and convivial way, as Latour and Schultz (2022) put it, through "envelopment" with the mother of us all—Nature. Let us aspire to working toward a goal that humanity's descendants get to experience it. First Nations and other indigenous peoples knew that once, and their inspiration leads us in reviving that hope.

References

Abate, Randall S. 2020. *Climate Change and the Voiceless: Protecting Future Generations, Wildlife, and Natural Resources.* Cambridge, UK: Cambridge University Press.
Abbott, Derek. 2012. Limits to Growth: Can Nuclear Power Supply the World's Needs. *Bulletin of Atomic Scientists,* 68 (5). https://thebulletin.org/2012/09/limits-to-growth-can-nuclear-power-supply-the-worlds-needs/.
Albrecht, Glenn. 2006. Environmental Distress as Solastalgia. *Alternatives,* 32 (4/5): 34–35.
Allen, Will. 2008. *The War on Bugs.* White River Junction, VT: Chelsea Green Publisher.
Alteri, Miguel. 2009. Agroecology, Small Farms and Food Sovereignty. *Monthly Review,* July 1: 34–43.
Ansar, Atif, Bent Flyvbjerg, Alexander Badzier, and Daniel Lunn. 2014. Should we Build More Large Dams? The Actual Costs of Hydropower Megaproject Development. *Energy Policy,* 69: 43–56.
Arkush, David and Donald Braman. 2023. Climate Homicide: Prosecuting Big Oil for Climate Deaths. *Harvard Environmental Law Review,* 48 (1) Available at SSRN: https://ssrn.com/abstract=4335779or http://dx.doi.org/10.2139/ssrn.4335779.
Baker, Janelle Marie. 2020. Bear Stories in the Berry Patch: Caring for Boreal Forest Fire Cycles. In *Extracting Oil in the Oil Sands: Settler Colonialism and Environmental Chage in Subarctic Canada.* Edited by Clinton Westman, Tara Joly, and Lena Gross, pp. 119–138. London, UK: Routledge.
Baker, Lauren E. 2008. Local Food Networks and Maize Agrodiversity Conservation: Two Case Studies from Mexico. *Local Environment,* 13 (3): 235–251.
Bakke, Gretchen. 2021. Pivoting Toward Energy Transition 2.0: Learning from Electricity. In *Research Handbook on Energy and Society.* Edited by James Webb, Faye Wade, and Margaret Tingey, pp. 97–111. Cheltenham, UK: Edward Elgar Publishing.
Bakke, Gretchen. 2016. *The Grid: The Fraying Wires Between American and Our Energy Future.* New York, NY: Bloomsbury.
Bates, Albert. 2010. *The Biochar Solution: Carbon Farming and Climate Change.* Gabriola Island, BC: New Society Publishers.
Bauwens, Michel, Visilis Kostakis, and Alex Pazaitis. 2019. *Peer to Peer: The Commons Manifesto.* London, UK: University of Westminster Press.
Bell, Shannon Elizabeth. 2009. "There Ain't No Bond in Town Like There Used to Be": The Destruction of Social Capital in the West Virginia Coal Mines. *Sociological Forum,* 24 (3): 631–657.

Bell, Shannon Elizabeth and Richard York. 2012. Coal, Injustice, and Environmental Destruction: Introduction to the Special Issue on Coal and the Environment. *Organization and Environment*, 24 (4): 359–368.

Bihouix, Philippe. 2014. *The Age of Low Tech: Towards a Technologically Sustainable Civilization*. Bristol, UK: Bristol University Press.

Blaikie, Piers. 1985. *The Political Economy of Land Erosion in Developing Countries*. London, UK: Longman.

Blaikie, Piers and Harold Brookfield. 1987. *Land Degradation and Society*. London, UK: Methuen.

Bodenhamer, Aysha and Tomas E. Shriver. 2020. Environmental Health Advocacy and Industry Obstruction: The Case of Black Lung Disease. *Rural Sociology*, 85 (3): 757–779.

Bollier, David. 2014. *Think Like a Commoner: A Short Introduction to the Life of the Commons*. Gabriola Island, BC: New Society Publishers.

Bollier, David and Silke Helfrich (Eds.). 2012. *The Wealth of the Commons: A World Beyond Market and State*. Amherst, MA: Levellers Press.

Bradford, Jason. 2019. *The Future is Rural: Food System Adaptations to the Great Simplification*. Corvallis OR: Post Carbon Institute.

Brown, Colin. 2015. Fungi Can Improve Agricultural Efficiency and Sustainability, *Yale Environment Review*. https://environment-review.yale.edu/authors/colin-brown#:~:text=Fungi%20can%20improve%20agricultural%20efficiency%20and%20sustainability%20A,also%20reducing%20the%20environmental%20impact%20of%20excess%20fertilizers.

Bryant, Raymond (Ed.). 2015. *The International Handbook of Political Ecology*. Cheltenham UK: Edward Elgar Publishing.

Bryant, Raymond. 1992. Political Ecology: An Emerging Research Agenda in Third World Studies. *Political Geography*, 11 (1): 12–36.

Burns, Shirley Stewart. 2007. *Bringing Down the Mountains: The Impact of Mountaintop Removal Surface Coal Mining on Southern West Virginia Counties*. Morgantown WVA: West Virginia University Press.

Carney, Mark. 2021. *Value(s): Building a Better World for All*. Toronto, ON: Penguin Random House Canada.

Carson, Rachel. 1962. *Silent Spring*. Boston, MA: Houghton Miflin.

Carter, Angela and Emily Eaton. 2010. Saskatchewan's Wild West Approach to Fracking. *The Monitor*. Ottawa, ON: Canadian Centre for Policy Alternatives (CCPA). https://www.policyalternatives.ca/publications/monitor/saskatchewan's-"wild-west"-approach-fracking.

Cernea, Michael. 1997. *Hydropower Dams and Social Impacts: A Sociological Perspective*. Social Development Papers. Paper No. 16. Washington, DC: The World Bank.

Cernea, Michael (Ed.). 1991. *Putting People First: Sociological Variables in Rural Development*, 2nd Edition, Revised and Expanded. New York: Oxford University Press for the World Bank.

Clark Michael A, Nina G. G. Domingo, Kimberly Colgan, Sumil Thakur, David Tilman, John Lynch, Ines L. Azevedo, and Jason D. Hill. 2020. Global Food System Emissions Could Preclude Achieving the 1.5 and 2 C Climate Change Targets. *Science*, 370: 705–708.

Coady, David, Ian Perry, Nigha-Piotr Le, and Baoping Shang. 2019. *Global Fossil Fuel Subsidies Remain Large: An Update on Country-Level Estimates*. Washington, DC: IMF Working Paper.

Coates, Ken. 2016. *First Nations Engagement in the Energy Sector in Western Canada*. Prepared for the Indian Resource Council. https://prosperitysaskatchewan.files.wordpress.com/2016/07/first-nations-engagement-in-the-energy-sector-in-western-canada.pdf.

Cordial, Paige, Ruth Riding-Molan, and Hillary Lios. 2012. The Effects of Mountaintop Removal Coal-Mining on Mental Health, Well-Being, and Community Health in Central Appalachia. *Ecopsychology*, 4 (3): 201–209.

Correia, Joel E. 2017. Soy States: Resource Politics, Violent Environments and Soybean Territorialization in Paraguay. *Journal of Peasant Studies*, 46 (2): 21–33.

Cox, Stan. 2020. *The Green New Deal and Beyond: Ending the Climate Crisis while We Still Can*. San Francisco, CA: City Lights Publishers.

Cox, Stan. 2009. Shrinking the Agricultural Economy Will Pay Big Dividends. *Synthesis/Regenesis* (48) http://www.greens.org/s-r/48/48-03.html

Cox, Stan and Priti Gulati Cox. 2022. Electric Vehicles Won't Save Us. *The Nation*. https://www.thenation.com/article/environment/electric-vehicles-lithium-cobalt-sustainable/.

Daly, Herman E. 2015. Economics for a Full World. *Great Transition Initiative*. https://greattransition.org/publication/economics-for-a-full-world.

Daly, Herman E. and Joshua Farley. 2011. *Ecological Economics: Principles and Applications*, 2nd Edition. Washington, DC: Island Press.

de Wit Chris (Ed.). 2005. *Development-Induced Displacement: Problems, Policies, and People*. Oxford, UK; Bergham.

De Young, Raymond and Thomas Prince (Eds.). 2012. *The Localization Reader*. Cambridge, MA: MIT Press.

Deemer, Bridget R., John A. Harrison, Siye Li, Jake J. Beaulieu, Tonya Delsontrom, Nathan Barros, Jose F. Bezerra-Neto, Stephen M. Powers, Marcos A. Dos Santos, and J. Arie Vonk. 2016. Greenhouse Gas Emissions from Reservoir Surfaces: A New Synthesis. *Bioscience*, 66 (11): 949–965.

DeHann, Lee, Marty Christians, Jared Crain, and Jesse Poland. 2018. Development and Evolution of an Intermediate Wheatgrass Domestication Program. *Sustainability*, 10. https://www.mdpi.com/2071-1050/10/5/1499.

Dembicki, Geoff. 2022. *The Petroleum Papers: Inside the Far-Right Conspiracy to Cover Up Climate Change*. Vancouver, BC: Greystone/David Suzuki Institute.

Deleuze, Gilles and Felix Guattari. 1987. *A Thousand Plateaus: Capitalism and Schizophrenia*. Minneapolis, MN: University of Minnesota Press.

Desmarais, Annette. 2007. *La Vía Campesina: Globalization and the Power of Peasants*. Halifax, NS: Fernwood Press.

Dowdall, Courtney Marie and Ryan J. Klotz. 2014. *Pesticides and Global Health: Understanding Agrochemical Dependence and Investing in Sustainable Solutions*. Walnut Creek, CA: Left Coast Press.

Duafala, A.P. 2018. The Historiography of the West Virginia Mine Wars. *West Virginia History*, 12 (1 & 2): 71–90.

Ehrenfield, David. 2002. Hard Times for Diversity. In *Fatal Harvest: The Tragedy of Industrial Agriculture*. Edited by Andrew Kimball, pp. 29–35. Washington, DC: Island Press.

Elliot, David. 2020. *Renewable Energy: Can it Deliver*. Cambridge, UK: Polity Press.
Encyclopedia of Saskatchewan. n.d. Hog Farming. https://esask.uregina.ca/entry/hog_farming.jsp.
Engdahl, F. William. 2007. *Seeds of Destruction: Hidden Agenda of Genetic Manipulation*. Montreal PQ: Global Research.
Epstein, Paul, Jonathan J. Buonocore, Kevin Eckerle, Michael Hendryx, Benjamin M. Stout III, Richard Heinberg, Richard W. Clapp, Beverly May, Nancy, L. Reinhart, Melissa M. Ahern, Samir K. Doshi, and Leslie Glustrom. 2011. Full Cost Accounting for the Life Cycle of Coal. *Annals of the New York Academy of Sciences*, 1219: 73–99.
Ervin, Alexander. 2015. *Cultural Transformations and Globalization: Theory, Development, and Social Change*. New York, NY: Routledge.
Ervin, Alexander. 2012. A Green Coalition Versus Big Uranium: Rhizomal Networks of Advocacy and Environmental Action. *Capitalism, Nature, Socialism*, 23 (3): 52–70.
Ervin, Alexander. 2011. The Vulnerabilities of Native People in the Mackenzie and Athabasca Drainage Systems: Tar Sands, Gas, and Uranium. In *Water, Cultural Diversity and Global Environmental Trends, Sustainable Futures?* Edited by Barbara Rose Johnston, pp. 277–291. Dordrecht, ND: UNESCO and Springer Publishing.
Ervin, Alexander. 2005. *Applied Anthropology: Tools and Perspectives for Contemporary Practice*, 2nd Edition. Boston, MA: Pearson.
Ervin, Alexander. 2001. *Canadian Perspectives in Cultural Anthropology*. Toronto, ON: Nelson/Thomson Learning.
Ervin, Alexander M., K. Holtzlander, D. Qualman, and R. Sawa (Eds). 2003. *Beyond Factory Farming: Corporate Hog Barns and the Threat to Public Health, the Environment, and Rural Communities*. Ottawa, ON: Canadian Centre for Policy Alternatives.
Escobar, Arturo. 2017. *Designs for the Pluriverse: Radical Interdependence, Autonomy, and the Making of Worlds*. Durham, NC: Duke University Press.
Escobar, Arturo. 2015. Degrowth, Postdevelopment, and Transitions: A Preliminary Conversation. *Sustainable Science*, 10: 451–462.
Escobar, Arturo. 1999. After Nature: Steps to an Anti-Essentialist Political Ecology. *Current Anthropology*, 40 (1): 1–30.
Escobar, Arturo. 1995. *Encountering Development: The Making and Unmaking of the Third World*. Princeton, NJ: Princeton University Press.
ETC Group. 2017. *Who Will Feed Us? The Industrial Food Chain vs The Peasant Food Web*, 3rd Edition. http://www.etcgroup.org/sites/www.etcgroup.org/files/files/etc-whowillfeedus-english-webshare.pdf.
Farley, Joshua. 2010. Ecological Economics. In *The Post Carbon Reader: Managing the 21st Century Sustainability Crises*. Edited by Richard Heinberg and Daniel Lerch, pp. 259–279. Heraldsburg, CA: Watershed Media.
Fiorino, Daniel J. 2018. *A Good Life on a Finite Earth: The Political Economy of Green Growth*. Oxford, UK: Oxford University Press.
Fitzgerald-Moore and P.J. Patel. 1996. *The Green Revolution*. Calgary, AB: University of Calgary. http://people.ucalgary.ca-pfitzger/green.pdf.
Fox, Julia. 1999. Mountaintop Removal in West Virginia: An Environmental Sacrifice Zone. *Organization and Environment*, 12 (2): 163–186.
Frank, Gunder. 1975. *On Capitalist Underdevelopment*. Oxford, UK: Oxford University Press.

Franzluebbers, Alan J., Laura K. Paine, Jonathan R. Winsten, Margaret Krome, Matt A. Sanderson, Kevin Ogles, and Dennis Thompson. 2012. Well-managed Grazing Systems: A Forgotten Hero of Conservation. *Journal of Soil and Water Conservation*, 67 (4): 100A–105A.

Gardiner, Stephen M. 2011. *A Perfect Moral Storm: The Ethical Tragedy of Climate Change*. Oxford, UK: Oxford University Press.

Georgescu-Roegen, Nicholas. 1971. *The Entropy Law and the Economic Process*. Cambridge, MA: Harvard University Press.

Gillies, Alexandra. 2020. *Crude Intentions: How Oil Corruption Contaminates the World*. Oxford, UK: Oxford University Press.

Gliessman, Stephen R., R. García, and M. Amador. 1981. The Ecological Basis for the Application of Traditional Agricultural Technology in the Management of Tropical Agrosystems. *Agro-Ecosystems*, 7: 173–185.

Glover, Jerry D. and John P. Reganold. 2010. Perennial Grains: Food Security for the Future. *Issues in Science and Technology*, 26: 1–47.

Gonzalez de Molina, Manuel, Paulo Frederico Petersen, Francisco Garrido Pena, and Franciso Roberto Caporal. 2020. *Political Agroecology: Advancing the Transition to Sustainable Food Systems*. New York:, NY CRC Press.

Greenberg, James and Thomas K. Park. 1994. Political Ecology: A Critical Introduction. *Journal of Political Ecology*, 1 (1): 1–12.

Griffith, Saul. 2021. *Electrify: An Optimist's Playbook for Our Clean Energy Future*. Cambridge, MA: MIT Press.

Gross, Lena. 2020. Wastelanding the Bodies, Wastelanding the Land: Accidents as Evidence in the Alberta Oil Sands. In *Extracting Oil in the Oil Sands: Settler Colonialism and Environmental Chage in Subarctic Canada*. Edited by Clinton Westman, Tara Joly, and Lena Gross. pp. 82–101. London, UK: Routledge.

Hansen, Arthur and Anthony Oliver-Smith. 1982. *Involuntary Migration and Resettlement: The Problems and Responses of Dislocated Peoples*. Boulder, CO: Westview Press.

Hansen, James. 2012. Coal: The Greatest Threat to Civilization. In *The Energy Reader: Overdevelopment and the Delusion of Endless Growth*. Edited by Tom Butler, Tom, Daniel Lerch and George Wuerthner, pp. 51–55. Sausalito, CA: Watershed Media.

Harding, Jim. 2007. *Canada's Deadly Secret: Saskatchewan Uranium and the Global Nuclear System*. Halifax,NS: Fernwood Publishing.

Hawken, Paul (Ed.). 2017. *Drawdown: The Most Comprehensive Plan Ever Proposed to Reduce Global Warming*. New York, NY: Penguin.

Hawken, Paul. 2007. *Blessed Unrest: How the Largest Movement in the World Came into Being and Why No One Saw it Coming*. New York, NY: Viking Press.

Hecht, Susanna. 1985. Environment, Development and Politics: Capital Accumulation and the Livestock Sector in Eastern Amazonia. *World Development*, 13 (6): 663–684.

Hecht, Susanna and Alexander Cockburn. 2010. *The Fate of the Forest: Developers, Destroyers, and Defenders of the Amazon*. Chicago, IL: University of Chicago Press.

Heinberg, Richard and Daniel Lerch (Eds.). 2010. *The Post Carbon Reader: Managing the 21st Century Sustainability Crise*s. Heraldsburg, CA: Watershed Media.

Hendriks, Richard, Phillip Raphals, Kare Bakker, and Gordon Christie. 2017. First Nations and Hydropower: The Case of British Columbia's Site C Dam Project. *Social Science Research Council, Items—Insights from the Social Sciences*. https://items.ssrc

.org/just-environments/first-nations-and-hydropower-the-case-of-british-columbias-site-c-dam-project/.
Heyneman, David. 1979. Dams and Diseases. *Human Nature*, 2 (2): 51–58.
Heyneman, David. 1971. Mis-aid to the Third World. *Canadian Journal of Public Health*, 63: 303–313.
Hickel, Jason. 2021. *Less is More: How Degrowth Will Save the World*. London, UK: Windmill Books.
Hickel, Jason. 2020. What Does Derowth Mean? A Few Points of Clarification. *Globalizations*. https//doi.org/10.1080/14747731.2020/812222.
Hickel, Jason. 2017. *The Divide: Global Inequality from Conquest to Free Markets*. New York, NY: W.W. Norton and Company.
Hoelle, Jeffrey. 2014. Cattle Culture in the Brazilian Amazon. *Human Organization*, 73 (4): 363–375.
Hoelle, Jeffrey. 2011. Convergence on Cattle: Political Ecology, Social Group Perceptions, and Socioeconomic Relationships in Acre Brazil. *Culture, Agriculture, Food and Environment*, 33 (2): 95–106.
Holmgren, David. 2002. *Permaculture: Principles and Pathways Beyond Sustainability*. Victoria, AU: Holmgren Design Services.
Hopkins, Rob. 2014. *From Oil Dependency to Local Resilience*, Illustrated Edition. Londo, UK: Chelsea Green Books.
Hornborg, Alf. 2016. *Global Magic: Technologies of Appropriation from Ancient Rome to Wall Street*. New York. N.Y.: Palgrave Macmillan.
Hornborg, Alf. 2015. Conceptualizing Ecologically Unequal Exchange: Society and Nature Entwined. In *The Routledge Handbook of Political Ecology*. Edited by Tom Perreault, Gavin Bridge, and James McCarthy, pp. 378–389. London, UK: Routledge.
Hornborg, Alf. 2011. *Global Ecology and Unequal Exchange: Fetishism in a Zero-Sum World*. London, UK: Routledge.
Hornborg, Alf. 2001. *Global Inequalities of Economy, Technology, and Environment*. Walnut Creek, CA: AltaMira.
Hornig, James F. (Ed.). 1999. *Social and Environmental Impacts of the James Bay Hydroelectric Project*. Montreal and Kingston, CA: McGill-Queens University.
Hossain, Eklas and Slobodan Petrovic. 2021. *Renewable Energy Crash Course: A Concise Introduction*. Cham, Switzerland: Springer.
Huseman, Jennifer and Damien Short. 2012. 'A Slow Industrial Genocide': Tar Sands and the Indigenous Peoples of Northern Alberta. *The International Journal of Human Rights*, 16 (1): 216–237.
Illich, Ivan. 1973. *Tools for Conviviality*. New York, NY: Harper and Row.
Jacobson, Mark Z. 2023. *No Miracles Needed: How Today's Technology Can Save Our Climate and Clean Our Air*. Cambridge, UK: Cambridge University Press.
Jacobson, Mark Z. 2020. *100% Clean, Renewable Energy and Storage for Everything*. Cambridge, UK: Cambridge University Press.
Jacobson, Mark Z. and Mark Delucci. 2009. A Path to Sustainable Energy by 2030. *Scientific American*, 301 (5): 58–66.
Jackson, Wes. 2010. *Consulting the Genius of the Place: An Ecological Approach to a New Agriculture*. Berkeley, CA: Counterpoint.
Jackson, Wes. 2002. Natural Systems Agriculture: A Truly Radical Agriculture. *Agriculture, Ecosystems and Environment*, 88 (2): 111–117.

Jackson, Wes and Rob Jenson. 2022. *An Inconvenient Apocalypse: Environmental Collapse, Climate Crisis, and the Fate of Humanity*. Notre Dame, IN: Notre Dame University Press.

Jencks, Clinton E. 1967. Social Status of Coal Miners Since Nationalization. *The American Journal of Economics and Sociology*, 26 (3): 301–312.

Johnston, Barbara Rose. 2009. Development Disaster, Reparations, and Right to Remedy: The Case of the Chixoy Dam in Guatemala. In *Development and Dispossession: The Crisis of Forced Displacement and Resettlement*. Edited by Anthony Oliver-Smith, pp. 201–225. Santa Fe, NM: School of Advanced Research Press.

Johnston, Barbara Rose (Ed.). 2007. *Half Lives and Half Truths: Confronting the Radioactive Legacies of the Cold War*. Santa Fe, NM: School for Advanced Research.

Johnston, Barbara Rose (Ed.). 1997. *Life and Death Matters: Human Rights and the Environment at the End of the Millennium*. Walnut Creek, CA: AltaMira.

Johnston, Barbara Rose (Ed.). 1994. *Who Pays the Price: The Sociocultural Context of Environmental Crisis* Washington, DC: Island Press.

Kallis, Giorgios, Frederico Demaria, and Giacomo D'Alisa. 2015. Introduction: Degrowth. In *Degrowth: A Vocabulary for a New Era*. Edited by D'Alisa, Giacomo, Fredrico Demaria, and Giorgos Kallis, pp. 1–25. New York, NY: Routledge.

Karl, Terry Lynn. 1999. The Perils of the Petro-State: Reflection on the Paradox of Plenty. *Journal of International Affairs*, 53 (1): 31–49.

Karl, Terry Lynn. 1997. *The Paradox of Plenty: Oil Boom and Petro-State*. Berkeley, CA: University of California Press.

Katusa, Marin. 2012. The Thing about Thorium: Why the Better Fuel. *Forbes*. https://www.forbes.com/sites/energysource/2012/02/16/the-thing-about-thorium-why-the-better-nuclear-fuel-may-not-get-a-chance/?sh=61711d371d80.

Kaviya, N., Vlabhav K. Upadhayay, Jyoti Singh, Amir Khan, Manisha Panwar, and Ajay Veer Singh. 2019. Role of Microorganisms in Soil Genesis and Functions. In *Mycorrhizosphere and Pedogenesis*. Edited by A. Varma and D.K. Choudhary, pp. 25–51. Singapore: Springer Nature.

Kedia, Satish. 2009. Health Consequences of Dam Construction and Involuntary Resettlement. In *Development and Dispossession: The Crisis of Forced Displacement and Resettlement*. Edited by Anthony Oliver-Smith, pp. 97–119. Santa Fe, NM: School of Advanced Research Press.

Kelton, Stephanie. 2020. *The Deficit Myth: Modern Monetary Theory and the Birth of the People's Economy*. New York, NY: Public Affairs.

Kimball, Andrew (Ed.). 2002a. *Fatal Harvest: The Tragedy of Industrial Agriculture*. Washington, DC: Island Press.

Kimball, Andrew. 2002b. Introduction. In *Fatal Harvest: The Tragedy of Industrial Agriculture*. Edited by Andrew Kimball, pp. 1–7. Washington, DC: Island Press.

Klein, Naomi. 2019. *On the Burning Case for A Green New Deal*. Toronto, ON: Knopf.

Klein, Seth. 2020. *A Good War: Mobilizing Canada for the Climate Emergency*. Toronto, ON: ECW Press.

Kneen, Brewster. 1995. *Invisible Giant: Cargill and Its Transnational Strategies*. Halifax, NS: Fernwood.

Kohr, Leopold. 2001. *The Breakdown of Nations* (Originally published 1957). Totnes, UK: Green Books.

Kolbert, Elizabeth. 2014. *The Sixth Extinction: An Unnatural History*. New York, NY: Henry Holt.
Kothari, Asish, Ariel Salleh, Arturo Escobar, Frederico Demaria, and Alberto Acosta (Eds.). 2019. *Pluriverse: A Post-Development Dictionary*. New Delhi, India: Tulika Books.
Kroese, Ron. 2002. Industrial Agriculture's War Against Nature. In *Fatal Harvest: The Tragedy of Industrial Agriculture*. Edited by Andrew Kimball, pp. 21–29. Washington, DC: Island Press.
Krieger, Tim and Martin Leroch. 2016. The Political Economy of Land Grabbing. *Homo Oecon*, 33: 197–204.
Lanker, Melissa, Michael Bell, and Valentin Picasso. 2019. Farmer Perspectives and Experiences Introducing the Novel Perennial Grain Kernza Intermediate Wheatgrass in the U.S. Midwest. *Renewable Agriculture and Food Systems*, 35 (6): 1–10.
Latour, Bruno. 2017. *Facing Gaia: Eight Lectures on the New Climatic Regime*. Cambridge, UK: Polity Press.
Latour, Bruno and Nikolaj Schultz. 2022. *On the Emergence of an Ecological Class—A Memo*. Cambridge, UK: Polity Press.
Leiserowitz, Anthony. 2020. Building Public and Political Will for Climate Change Action. *Yale School of the Environment News,* June 30. https://environment.yale.edu/news/article/building-public-and-political-will-for-climate-change-action.
Leiserowitz, Anthony, Connie Roser-Renouf, Jennifer Marlon, and Edward Maibach. 2021. Global Warming's Six Americas and Recommendations for Climate Change Communication. *Current Opinion in Behavioral Sciences,* 42: 97–103.
Lerer, Leonard B. and Thayer Scudder. 1999. Health Impacts of Large Dams. *Environmental Assessment Review*, 19: 113–123.
Lewis, Simon L. and Mark Maslin. 2015. Defining the Anthropocene. *Nature*, 519: 171–180.
Longley, Hereward. 2020. Uncertain Sovereignty: Treaty 8. Bitumen and Land Claims in the Athabasca Oil Sands Region. In *Extracting Oil in the Oil Sands: Settler Colonialism and Environmental Chage in Subarctic Canada*. Edited by Clinton Westman, Tara Joly, and Lena Gross, pp. 23–48. London, UK: Routledge.
Lopes de Souza, Marcello. 2021. 'Sacrifice Zone': The Environment—Territory—Place of Disposable Lives. *Community Development Journal*, 56 (2): 220–243.
Lovelock, James. 1979. *Gaia: A New Look at Life on Earth*. Oxford, UK: Oxford University Press.
Lovins, Amory B. 1996. Negawatts: Twelve Transitions, Eight Improvements, and One Distraction. *Energy Policy,* 24 (4): 331–343.
Lovins, Amory B. 1990. The Negawatt Revolution. *Across the Board* (The Conference Board Magazine), 27 (9): 18–24.
Lovins, Amory B. and Rocky Mountain Institute. 2011. *Reinventing Fire: Bold Business Solutions for the New Energy Era*. White River Junction, VT: Chelsea Green Publishing.
MacKay, David J.C. 2009. *Sustainable Energy without the Hot Air*. Cambridge, UK: UIT Limited.
MacVittie, Susan. 2013. Land Grabs. *Watershed Sentinel*. March-April, 26–29.
Malm, Andreas. 2021. *How to Blow Up a Pipeline*. New York, NY: Verso.
Mander, Jerry. 2002. Machine Logic: Industrializing Nature and Agriculture. In *Fatal Harvest: The Tragedy of Industrial Agriculture*. Edited by Andrew Kimball, pp. 17–21. Washington, DC: Island Press.

Mann, Geoff and Joel Wainwright. 2018. *Climate Leviathan: A Political Theory of Our Planetary Future*. London, UK: Verso.

Mann, Michael E. 2021. *The New Climate War: The Fight to Take Back our Planet*. London, UK: Scribe.

Manning, Richard. 2004. *Against the Grain: How Agriculture Has Hijacked Civilization*. New York: North Point Press.

Marberry, M. Katie and Danilea Werner. 2020. The Role of Mountaintop Removal Mining in the Opioid Crisis. *Journal of Social Work Practice in the Addictions*, 29 (4): 302–310.

Mayer, Jane. 2017. *Dark Money and the Hidden History of the Billionaires Behind the Rise of the Radical Right*. New York, NY: Doubleday.

McCay, Bonnie and James Acheson (Eds.). 1987. *The Question of the Commons: The Culture and Ecology of Communal Resources*. Tucson, AZ: University of Arizona Press.

McCormack, Patricia A. 2020. Conclusion: Studying the Social and Cultural Impacts of "Extreme Extraction" in Northern Alberta. In *Extracting Oil in the Oil Sands: Settler Colonialism and Environmental Chage in Subarctic Canada*. Edited by Clinton Westman, Tara Joly, and Lena Gross, pp. 180–199. London, UK: Routledge.

McCusker, Brent. 2015. Political Ecology and Policy: A Case Study in Engagement. In *The Routledge Handbook of Political Ecology*. Edited by Tom Perreault, Gavin Bridge, and James McCarthy, pp. 188–198. London, UK: Routledge.

McKibben, Bill. 2010. *eaarth: Making Life on a Tough New Planet*. New York, NY: Holt, Rinehart, and Winston.

Méndez, A. Ernesto, Christopher M. Bacon, and Roseann Cohen (Eds.). 2016. *Agroecology: A Transdisciplinary, Participatory, and Action-Oriented Approach*. Boca Raton, FL: CRC Press.

Merchant, Carolyn. 1980. *The Death of Nature: Women, Ecology and the Scientific Revolution*. New York, NY: HarperOne.

Mighty Earth. 2017. *Mystery Meat II: The Industry Behind the Quiet Destruction of the American Heartland*. http://www.mightyearth.org/wp-content/uploads/2017/08/Meat-Pollution-in-America.pdf. Washington, DC.: Center for International Policy.

MIT (Massachusetts Institute of Technology). 2006. *Project Amazonia: Threats—Agriculture and Cattle Ranching*. http://web.mit.edu/12.000/www/m2006/final/threats/threat_agg.html.

Monbiot, George. 2022. *Regenesis Feeding the World Without Devouring the Planet*. London, UK: Allen Lane.

Moran, Emilio F., Maria Claudio Lopez, Nathan Moore, Norbert Mulller, and David W. Hyndman. 2018. Sustainable Hydropower in the 20th Century. *Papers of the National Academy of Sciences (PNAS)*, 115 (47): 11891–11898.

Morrice, Emily and Ruth Colagiuri. 2013. Coal Mining, Social Injustice and Health: A Universal Conflict of Power and Priorities. *Health and Place*, 19: 74–79.

Nida, Brandon. 2013. Demystifying the Hidden Hand: Capital and the State at Blair Mountain. *Historical Archaeology*, 47 (3): 32–68.

Nikiforuk, Andrew. 2023. The Rising Chorus of Renewable Energy Skeptics. *The Tyee*. https://thetyee.ca/Analysis/2023/04/07/Rising-Chorus-Renewable-Energy-Skeptics/.

Nikiforuk, Andrew. 2010. Dirty Oil is Turning Canada into a Corrupt Petro-State. *The CCPA Monitor*, May: 10–11.

Nikiforuk, Andrew. 2008. *Tar Sands: Dirty Oil and the Future of a Continent*. Vancouver, BC: Greystone Books.

Norberg-Hodge, Helena. 2019. *Local is Our Future: Steps Toward an Economics of Happiness*. White River Junction, VT: Chelsea Green Press..

Norberg-Hodge, Helena. 2002. Global Monoculture: The Worldwide Destruction of Diversity. In *Fatal Harvest: The Tragedy of Industrial Agriculture*. Edited by Andrew Kimball, pp. 13–17. Washington, DC: Island Press.

Nouh, Fatma Ahmed Abo, Hebatallah H. Abo Nahas, and Ahmed M. Abdel-Azeem. 2020. Agriculturally Important Fungi: Plant-Microbe Association for Mutual Benefits. In *Fungi for Sustainable Agriculture*. Edited by A.N. Yadav et al., pp. 1–13. Cham, Switzerland: Springer Nature.

Office for Economic Cooperation and Development (OECD). 2011. *Towards Green Growth: A Summary for Policy Makers*. Online https://www.oecd.org/greengrowth/48012345.pdf.

Oliver-Smith, Anthony (Ed.). 2009a. *Development and Dispossession: The Crisis of Forced Displacement and Resettlement*. Santa Fe, NM: School of Advanced Research Press.

Oliver-Smith, Anthony. 2009b. Introduction: Development-Forced Displacement and Resettlement: A Global Human Rights Crisis. In *Development and Dispossession: The Crisis of Forced Displacement and Resettlement*. Edited by Anthony Oliver-Smith, pp. 3–25. Santa Fe, NM: School of Advanced Research Press.

Oreskes, Naomi and Erik Conway. 2011. *Merchants of Doubt: How a Handful of Scientists Obscured the Truth on Issues from Tobacco Smoke to Global Warming*. New York, NY: Bloomsbury.

Orrego, Juan Pablo. 2012. River Killers the False Solution. In *The Energy Reader: Overdevelopment and the Delusion of Endless Growth*. Edited by Tom Butler, Daniel Lerch, and George Wuerthner, pp. 179–186. Sausalito, CA: Watershed Media.

Ostrom, Elinor. 1990. *Governing the Commons: The Evolution of Institutions for Collective Action*. Cambridge, UK: Cambridge University Press.

Patel, Raj. 2009a. *The Value of Nothing: Why Everything Costs So Much More than We Think*. Toronto, ON: Harper Collins.

Patel, Raj. 2009b. What Does Food Sovereignty Look Like? *The Journal of Peasant Studies*, 36 (3): 663–706.

Patel, Raj. 2007. *Stuffed and Starved: The Hidden Battle for the World's Food System*. Toronto, ON: Harper Perennial Press.

Peake, Stephen. 2021. *Renewable Energy: Ten Short Lessons*. London, UK: Michael O'Mara Books.

Perreault, Tom, Gavin Bridge, and James McCarthy (Eds.). 2015. *The Routledge Handbook of Political Ecology*. London, UK: Routledge.

Perrins, Dan. 2009. *Future of Uranium Consultation Process*. Regina SK: Report for the Saskatchewan Government.

Pimental, David and Michael Burgess. 2013. Soil Erosion Threatens Food Production. *Agriculture*, 3: 443–463.

Pollan, Michael. 2006. *The Omnivore's Dilemma: A Natural History of Four Meals*. New York, NY: The Penguin Press.

Raworth, Kate. 2017. *Doughnut Economics: 7 Ways to Think Like a 21st Century Economist*. White River Jt., VT: Chelsea Green.

Rees, William. 2014. Avoiding Collapse: An Agenda for Sustaianble Degrowth and Relocalization of the Economy. Canadian Centre for Policy Alternatives. ccpa-bc AvoidingCollapse Rees.pdf (policyalternatives.ca)

Reyna, Stephen and Andrea Behrends. 2008. The Crazy Curse and Crude Domination: Toward an Anthropology of Oil. *Focall—European Journal of Anthropology*, 52: 3–17.

Rhodes, Richard. 2018. *Energy: A Human History*. New York, NY: Simon and Schuster.

Rifkin, Jeremy. 2019. *The Green New Deal: Why the Fossil Fuel Civilization Will Collapse by 2028 and The Bold Economic Plan to Save Life on Earth*: New York, NY: St Martin's Press.

Rifkin, Jeremy. 1992. *Beyond Beef: The Rise and Fall of the Cattle Culture*. New York, NY: Plume.

Robbins, Paul. 2019. *Political Ecology*, 3rd Edition. New York, NY: John Wiley and Sons.

Robbins, Richard. 2011. *Global Problems and the Culture of Capitalism*, 5th Edition. Upper Saddle River, MA: Prentice Hall.

Robin, Monique. 2010. *The World according to Monsanto: Pollution, Corruption and the Control of Our Food System*. Cambridge, UK: Cambridge University Press.

Rosaldo-May, Francisco J. 2016. The Intercultural Origins of Agroecology: The Contributions from Mexico. In *Agroecology: A Transdisciplinary Participatory and Action-Oriented Approach*. Edited by V. Ernesto Méndez, Christopher Bacon, and Roseann Cohen., pp. 123–139. Boca Raton, FL: CRC Press.

Rosset, Peter and Medea Benjamin. 1994. *The Greening of the Revolution: Cuba's Experiment with Organic Agriculture*. Melbourne, AU: Ocean Press.

Sahlins, Marshall. 1972. *Stone Age Economics*. New York, NY: de Gruyter.

Sawyer, S. 2004. *Crude Chronicles: Indigenous Politics, Multinational Oil, and Neoliberalism in Ecuador*. Durham, NC: Duke University Press.

Scheidler, Fabian. 2020. *The End of the Megamachine: A Brief History of a Failing Civilization*. Winchester, UK: Zero Books.

Schumacher, E.F. 1974. *Small is Beautiful: A Study of Economics as if People Mattered*. London, UK: Abacus.

Schumpeter, Joseph. 1942. *Capitalism, Socialism, and Democracy*. New York, NY: Haper and Brothers.

Scott, James C. 1998. *Seeing Like a State: How Certain Schemes to Improve the Human Condition Have Failed*. New Haven, CT: Yale University Press.

Scranton, Roy. 2015. *Learning to Die in the Anthropocene: Reflections on the End of a Civilization*. San Francisco, CA: City Lights Publishers.

Scudder, Thayer. 2019. *Large Dams: Long Term Impacts on Riverine Communities and Free Flowing Rivers*. Singapore: Springer Publishing.

Scudder, Thayer. 2005. *The Future of Large Dams: Dealing with Social, Environmental, Institutional, and Political Costs*. London, UK: Earthscan.

Sheldrake, Merlin. 2020. *Entangled Life: How Fungi Make Our Worlds, Change Our Minds, and Shape Our Futures*. New York, NY: Random House.

Shiva, Vandana. 1997. *Biopiracy: The Plunder of Nature and Knowledge*. Cambridge, MA: Southend Press.

Singh, Arun Kumar. 2020. Development Induced Displacement: Issues and Indian Experiences. *Journal of the Anthropological Survey of India*, 69 (2): 276–289.

Smaje, Chris. 2020. *A Small Farm Future: Making the Case for a Society Built around Local Economies. Self-Provisioning Agricultural Diversity, and a Shared Earth.* White River Junction, VT: Chelsea Green Publishing.

Smil, Vaclav. 2019. Growth from Microorganisms to Megacities. Boston, MA: MIT Press.

Smil, Vaclav. 2017. *Energy and Civilization: A History.* Boston, MA: MIT Press.

Smil, Vaclav. 2014. A Global Transition to Renewable Energy Will Take Many Decades. *Scientific American.* https://www.scientificamerican.com/article/a-global-transition-to-renewable-energy-will-take-many-decades/.

Snyder, Nancy. 2023. The Climate Crisis Goes to Court; Youth Prevails. *CounterPunch.* https://www.counterpunch.org/2023/08/25/the-climate-crisis-goes-to-court-youth-prevails/.

Soleri, Daniela and David Arthur Cleaveland. 2006. Transgenic Maize and Mexican Maize Diversity. *Agriculture and Human Values*, 23 (1): 27–31.

Soto-Gómez and Paula Pérez-Rodríguez. 2022. Sustainable Agriculture through Perennial Grains: Wheat, Rice, Maize, and Other Species: A Review. *Agriculture, Ecosystems, and Environment*, 325: 107747. https://www.sciencedirect.com/science/article/pii/S0167880921004515.

Specht, Joshua. 2019. *Red Meat Republic: A Hoof-to-Table History of How Beef Changed.* America. Princeton, NJ: Princeton University Press.

Steinfield, A., H.A. Monney, L.E. Neville, P. Garber and R. Reid. 2008. *Livestock in a Changing Landscape.* U.N.E.S.C.O. Policy Brief. https://unesdoc.unesco.org/ark:/48223/pf0000159194.

Stoffle, Richard W. and Michael Evans. 1990. Holistic Conservation and Cultural Triage: Indian Perspectives on Cultural Resources. *Human Organization*, 49 (2): 91–99.

Thierry, Wim, Stefan Lange, Joeri Rogelj, Carl-Freidrich Schleussner, Lukas Gudmundsson, Sonia I. Seneviratne, Marina Andrijevic, Katja Frieler, Kerry Emmanuel, Tobias Geiger, David N. Bresch, Fang Zhao, Sven N. Willner, Matthias Büchner, Jan Volholtz, Nico Bauer, Jinfeng Chang, Phillipe Ciais, Marie Dury, Louis François, Mannolis Grillakis, Simon. N. Gosling, Naota Hanask, Thomas Hickler, Veronika Huber, Akihiko Ito, Jona Jägermeyer, Nikolay Khabarov, Aristeidis Koutroulis, Wengfeng Liu, Wolfgang Lutz, Matthias Mengel, Christoph Müller, Sebastian Ostberg, Christopher P.O. Reyer, Tobias Stacke, and Yoshida Wada. 2021. Intergenerational Inequities in Exposure to Climate Extremes. *Science*, 374. (6564): 158–161.

Thompson, Edward Palmer. 1991.*The Making of the English Working Class* (Originally published 1963). London, UK: Penguin.

Thu, Kendall. 2010. CAFOS are in Everybody's Backyard: Industrial Agriculture, Democracy and the Future. In *The CAFO: Reader: The Tragedy of Industrial Animal Factories.* Edited by Daniel Imhoff, pp. 210–221. Berkeley CA: University of California Press.

Thu, Kendall and Paul Durrenberger. 1998. *Pigs, Profits, and Rural Communities.* Albany, NY: State University of New York.

Thunberg, Greta (Ed.). 2023. *The Climate Book.* New York, NY: Penguin.

Timoney, Kevin P. and Peter Lee. 2009. Does the Alberta Tar Sands Industry Pollute? The Scientific Evidence. T*he Open Conversation Biology Journal* (3): 65–81.

Touraine, Alain. 1985. An Introduction to a Study of Social Movements. *Social Research*, 52: 749–788.
Uranium Development Partnership. 2009. *Capturing the Full Potential of the Uranium Value Chain in Saskatchewan*. Regina, SK: Report for the Saskatchewan Government.
Urquhart, Ian. 2018. *Costly Fix: Power, Politics, and Nature in the Tar Sands*. Toronto, ON: University of Toronto Press.
Vallianatos, E.G. 2014. *Poison Spring: The Secret History of Pollution and the EPA*. New York, NY: Bloomsbury Pr
Van Genuchten, Erlijn. 2022. How Fungi Can Help Plants Grow and Make Agriculture More Sustainable: 5 Reasons Fungi Can Replace Chemicals in Making Plants Thrive. Sharing Science: https://medium.com/a-microbiome-scientist-at-large/how-fungi-can-help-plants-grow-and-make-agriculture-more-sustainable-bc126a1b7019.
Varoufakis, Yanis. 2017. *Talking to My Daughter About the Economy or, How Capitalism Works—and How It Fails*. New York, NY: Farrar, Straus, and Giroux.
Vetter, Andrea. 2018. The Matrix of Convivial Technology—Assessing Technologies for Degrowth. *Journal of Cleaner Production*, 197: 1776–1785.
Vicedo-Cabrera, A.M., N. Scovronick, F. Sera, D. Royé, R. Schneider, A. Tobias, C. Astrom, Y. Guo, Y. Honda, D.M. Hondula, R. Abrutzky, S. Tong, M. de Sousa Zanotti Stagliorio Coelho, P.H. Nascimento Saldiva, E. Lavigne, P. Matus Correa, N. Valdes Ortega, H. Kan, S. Osorio, J. Kyselý, A. Urban, H. Orru, E. Indermitte, J.J.K. Jaakkola, N. Ryti, M. Pascal, A. Schneider, K. Katsouyanni, E. Samoli, F. Mayvaneh, A. Entezari, P. Goodman, A. Zeka, P. Michelozzi, F. de'Donato, M. Hashizume, B. Alahmad, M. Hurtado Diaz, C. De La Cruz Valencia, A. Overcenco, D. Houthuijs, C. Ameling, S. Rao, F. Di Ruscio, G. Carrasco-Escobar, X. Seposo, S. Silva, J. Madureira, I.H. Holobaca, S. Fratianni, F. Acquaotta, H. Kim, W. Lee, C. Iniguez, B. Forsberg, M.S. Ragettli, Y.L.L. Guo, B.Y. Chen, S. Li, B. Armstrong, A. Aleman, A. Zanobetti, J. Schwartz, T.N. Dang, D.V. Dung, N. Gillett, A. Haines, M. Mengel, V. Huber, and A. Gasparrini. 2021. The Burden of Heat-Related Mortality Attributable to Recent Human-Induced Climate Change. *Nature Climate Change*, 11: 492–500.
Voyles, Traci Byrnne. 2015. *Wastelanding: Legacies of Uranium Mining in Navajo Country*. Minneapolis, MN: University of Minnesota Press.
Wackernagel, M. and W.E. Rees. 1996. *Our Ecological Footprint*. Gabriola Island, BC: New Society Publishers.
Walker, Nathalie, Sabrina Patel, and Kemel Kalif. 2013. From Amazon Pasture to the High Street: Deforestation and the Brazilian Cattle Product Supply Chain. *Tropical Conservation Science Special Issue*, 6 (3): 446–467.
Wallerstein, Immanuel. 2004. *World Systems Analysis.: An Introduction*. Durham, NC: Duke University Press.
Wang, Ming Xiao, Zhang Tao, Xie Miao-Rong, Zhang Bin, and Jia, Ming-Qui. 2011. Analysis of National Coal Mining Accident Data in China, 2001–2008. *Public Health Reports*, 126: 270–276.
Watts, Laura. 2018. *Energy at the End of the World: An Orkney Island Saga*. Cambridge, MA: The MIT Press.
Watts, Michael. 2015. Now and Then: The Origins of Political Ecology and the Rebirth of Adaptation as Form of Thought. In *The Routledge Handbook of Political Ecology*. Edited by Tom Perreault, Gavin Bridge, and James McCarthy, pp. 19–51. London, UK: Routledge.

Watts, Michael. 2004a. Resource Curse? Governmentality, Oil, and Power in the Niger Delta Nigeria. *Geopolitics*, 9 (1): 50–80.

Watts, Michael. 2004b. Violent Environments: Petroleum Conflict and the Political Ecology of Rule In *Liberation Ecologies: Environment, Development, Social Movements*, 2nd Edition. Edited by Richard Peet and Michael Watts, pp. 273–299. London, UK: Routledge.

Watts, Michael. 1983. *Silent Violence: Food, Famine and Peasantry in Northern Nigeria*. Berkeley, CA: University of California Press.

Westman, Clinton. 2013. Social Impact Assessment and the Anthropology of the Future in Canada's Tar Sands. *Human Organization*, 72 (2): 111–120.

Westman, Clinton and Tara Joly. 2019. Oil Sands Extraction in Alberta, Canada, a Review of impacts and Processes Concerning Indigenous Peoples. *Human Ecology*, 47: 233–243.

Westman. Clinton, Tara Joly, and Lena Gross (Eds.). 2020a. *Extracting Oil in the Oil Sands: Settler Colonialism and Environmental Chage in Subarctic Canada*. London, UK: Routledge.

Westman, Clinton, Lena Gross, and Tara Joly. 2020b. Introduction. In *Extracting Oil in the Oil Sands: Settler Colonialism and Environmental Change in Subarctic Canada*. Edited by Clinton Westman et al., pp. 1–23. London, UK: Routledge.

Wittman, Hannah, Annette Desmarais, and Nettie Wiebe (Eds.). 2010. *Food Sovereignty: Reconnecting Food, Nature, and Community*. Halifax, NS: Fernwood Press.

Wolf, Eric. 1982. *Europe and the People Without History*. Berkeley, CA: University of California Press.

World Commission on Dams. 2000. *Dams and Development. A New Framework for Decision-making*. The Report of the World Commission on Dams. London, UK: Earthscan.

Wright, Tim. 2013. Coal Mining in China: The Social Costs. *Culture and Society*, 14 (1): 143–152.

Index

4th Industrial Revolution 145
350.org 8, 219
2008 Global Financial Crisis 70, 160

Adam, Chief Allan 106
advertising, reducing 154
advocacy coalitions 220
AECL *see* Atomic Energy of Canada Corporation Limited
Africa, children born in 2020 210
agribusiness 23, 28, 70
agriculture 13–16, 27–28; agroecology 163–169; annual crops 175–176; biotechnology 19; critique of industrial agriculture 16–20; diversity 17–19, 164, 167–168; energy costs 74; environmental concerns 29–30; fertilizers 169; foreign land grabs 26–27; Green Revolution 20–22; high-yielding varieties (HYVs) 21, 23; methane 30; monocrop systems 16, 176, 226; natural systems agriculture 175–180; perennial grains 175–180; permaculture 180; plant breeding 16, 48, 179; polyculture systems 167–168; regenerative or restorative agriculture 169–175; urban agriculture 66; value chains 25–26; water 21–22; *see also* fertilizers; field crops
agroecology 64, 163–169
air conditioning 199
Alarmed people 219–220
Amazonian soils 172
American buffalo 49–55
American New Deal 223
ammonia nitrate 19–20
amoralism 18
annual crops 175–176

Anthropocene 1
Archer Daniels Midland (ADM) 36, 42, 44, 49, 64
artificial fertilizers 17, 19–20; *see also* fertilizers
Aswan High Dam 112
Athabasca Basin 125
Athabasca River 96
Athabasca Tar Sand Healing Walks 106–107
atmospheric colonialism 153
Atomic Energy of Canada Corporation Limited (AECL) 119–120, 127
"Atoms for Peace" Program 117
avian flus 67

Bakke, Gretchen 198, 203
batteries: electric vehicles 201; storing 196–197
Bayer/Monsanto 36, 42, 44, 46, 64, 73
beef industry 155; Brazil 55–57; rise of 49–55
biochar solution 172–173
bioenergy 190–191
biogeochemistry 61
biomass 173, 191
biotechnology in agriculture 19
bitumen 1, 96, 102
black lung disease 89
Blair Mountain Battle 85; National Historic Site 89
blue baby alerts 34
Brazil: agriculture 22, 24; cattle ranching 55–57; soy 44–47
Britain: beef industry 51–52; coal 80; nuclear reactors, decommissioning 134
British Petroleum 91
Bruce Power 120, 122

buck-passing 207–208
buildings, and renewable energy 194–195
Bunge 49
Butz, Earl 35

CAFOs *see* confined animal feeding operations
Cameco 121
Campbell wheat farm 15
Canada: Athabasca Tar Sand Healing Walks 106–107; controlled fires 101; cost of cleaning oil operations 103–104; environmental impact assessment (EIA) 98; First Nations *see* First Nations; Green New Deal 161; hydroelectric power 115; indigenous peoples 99–102; nuclear energy 118; petroleum 94; search for nuclear repository 131–134; tar sands 94–104; uranium mining 119–120; World War II 146
Canadian Nuclear Waste Management Organization (NWMO) 127–128
Canadian Shield 127
CANDU heavy-water cooled reactors 118, 127
cap and adapt 223
cap and trade systems 144
capitalism 141, 144, 157–158, 160; carbon-fueled capitalism 222; corporate capitalism 161; critique of 152; mission-oriented capitalism 161; monopoly capitalism 28
capitalist modernity, critique of 150
carbon capture technologies 182
carbon dioxide emissions 111, 153, 213; coal 81; reducing 224
carbon-fueled capitalism 222
carbon taxes 144–145
Cargill of Minneapolis 22, 24, 36, 42, 44, 49, 64
Carney, Mark 143–144, 160
cattle ranching, Brazil 55–57
Cautious people 219
CCGS *see* Coalition for a Clean Green Saskatchewan
cereals 37
Chalk River, uranium mining 119
Chernobyl 124, 134
Chevron 91, 214
children born in 2020 210

China: Climate Mao 158–159, 230; coal 82–83; hydro power 189
chinampas (raised garden) 166
Chixoy Dam 109
city states 225
civilization collapse, potential response to 225–229
class consciousness 217
class struggle 217
Climate Behemoth 157, 159
climate change 31, 72
Climate Change Communication 219
Climate Corps 148
climate events 210, 225
Climate Leviathan 157, 159, 230
Climate Mao 158–159, 230
Climate X 158–159, 215–221
CO_2 emissions 142–143; *see also* carbon dioxide emissions
coal 79–80, 89–90, 200; China 82–83; environmental degradation 82; mining 81; mountain-top removal coal mining (West Virginia) 83–90
coal ash 82
Coalition for a Clean Green Saskatchewan (CCGS) 124–125, 127, 129–130, 133, 215
cobalt 201
Cold War era 118
colonialism, atmospheric colonialism 153
Colorado River 111
Committee for Future Generations (Saskatchewan) 133–134
commons, reclaiming 228
compressed air 197
concentrated solar power (CSP) 185
Concerned people 219
concrete production, carbon dioxide emissions 111
Conferences of the Parties (COPs) 157–158
confined animal feeding operations (CAFOs) 27, 36; pork production 57–60
Conservation Corps 148
contemporary agriculture 13–16
continuous economic growth 141
controlled fires, Canada 101
conviviality in low-tech society 227–228
cooperative behavior 228

cooperatives 227–228
Copenhagen Accord (2012) 157
copper 202
COPs *see* Conferences of the Parties
corn 32–38; North American Free Trade Agreement 39–43
corn whiskey 37–38
corporate capitalism 161
corporate massification 19
cost: of cleaning up oil operations 103–104; of cleaning uranium mines (Saskatchewan) 121–122; of energy 201; of nuclear reactors 134
cost overruns, dam constructions 111
cost-squeeze pressures 26, 29
courts: First Judicial District Court of Montana 214; rulings on climate change 212, 214
COVID-19 pandemic 160; suspension of environmental impact inspections (Canada) 98
Cox, Stan 201, 223
crazy curse of oil 94
creative destruction 184
Cree people, Canada 101–103
crop disease 16
crop rotations 166
CSP *see* concentrated solar power
Cuba: agroecology 165; sufficiency 226

Daly, Herman 150–151
dam constructions 109
dams, hydroelectric dams 108–116
decommissioning nuclear reactors 135–137
decomposing biomass 173
degrowth 141, 149–156, 159, 161–162, 223, 229
Dene people 132–133
Denmark, wind power 186
depopulation, North American communities 29
desertification 61
destructive agency 1; scaling down 155
development 150; without significant growth 153
development forced displacement (DFD) 110
Disengaged people 219
disinformation 213
Dismissive people 219–220

diversity in agriculture 17–19, 164, 167–168
Dounreay nuclear facility 135
downriver impacts, from dams 112
downstream communities 113
drag-lining coal 86
Dutch disease 92
dysbiosis 171

earthworms 169–170
ecological footprints 226
economic models 229
EIA *see* environmental impact assessment
Eisenhower, President Dwight 118
ejido (village ownership of commons of land with shared labor and produce) 165
Eldorado 119
electrical grid, redesigning 198–200
electric vehicles 145, 192–194, 201
Elliot Lake, Ontario, uranium mining 119
ending food waste 155
endophytic fungi 171
energy: coal *see* coal; hydroelectric dams 108–116; nuclear energy *see* nuclear energy; petroleum *see* petroleum
energy consumption 2; agriculture 74
energy efficiency 200
English River First Nation, Canada 133
envelopment 218
environmental concerns, agriculture 29–30
environmental degradation: from coal 82; industrial farming 70; from tar sands (Canada) 97–98
environmental impact assessment (EIA), Canada 98
environmental impact studies 6–7
environmental laws 5
Escobar, Arturo 4, 151–152
externalities 160
extractivism 91
ExxonMobil 91, 214

family farms 15
farmer protests 64
farmers 13–14, 28–29; agroecology 167; in India 63; managed paddock grazing 174–175; small-farmer food webs 65–68; small-scale farmers 166–167

farming: agroecology 164–169; industrial farming 63
farms 13–16
fertilizers 15, 30, 33–34, 67, 169; artificial fertilizers 17, 19–20; nitrogen fertilizer 23
field crops 47–48; corn 32–38; soy 43–47
financial mechanisms 143
First Judicial District Court of Montana 214
First Nations, Canada 95, 99, 103, 132; opposition to petroleum projects 104; participation in tar sands development 104–105; uranium mining 119–120
Fishing, damaged 113
flooding: from coal mining 86; due to dams 111
fly ash, coal 82
food, government control of 63–64
food insecurity 30, 173
food policies 64
food preservation 37
food prices 34–35
food processing companies 24–25
food processors 28
food security 17, 63, 70–71
food sovereignty 64, 163, 164; La Via Campesina (LVC) 68–72
food systems 164
food waste 155, 180
foreign land grabs 26–27
Fort Chipewyan 96
fossil fuels 183; coal *see* coal; industrial farming 67; oil *see* petroleum
fracking 5, 213
fractional reserve banking systems 151
France, nuclear energy 118
free trade 25
Fridays for Future climate strikes 211
Fukushima disaster 134
fungi 170–171
future generations 208–210; court rulings protecting 214; legislative impact on 212
Future of Uranium Commission 125
Future of Uranium report 129–130

Gaia 7
Gardiner, Stephen 207–210
General Foods 24
General Mills 36, 179

genetically modified organisms (GMOs) 16, 19, 67, 73; corn 39, 42; maize 66; soy 46
geothermal energy 189–190
Germany, solar energy 203
germplasm 165
global governance systems 151
globalization 18; of production 168
global monoculture 17
Global North 141, 218
global seed market 67
Global South 141, 153, 216
global temperatures 144
global warming, potential response to 225–229
GMOs *see* genetically modified organisms
government debt 160
governments: control of food 63–64; degrowth 223; Green New Deal 145–149; lobbying governments 213; protecting future generations 208; regulatory policies 144
grains 47
Great Bear Lake 119
great disconnect 16, 28
Green Corps 148
greenfield status 134–135
Green Growth 141–145, 159
greenhouse gas emissions 180
Green New Deals 141, 145–149, 159, 222
Greenpeace 8, 105
Green Revolution 20–22, 24, 29; political motives for 22–25
Green Revolution Two 70, 72
Griffith, Saul 182–183
Gulf States 93
Gunnar Uranium Mine 104, 121

Haber-Bosch process 33–34
habitability for all 218
hamburgers 53–54
health of people: impact of coal 86–87; impacts of dams 112; near tar sands (Canada) 96–97; uranium mining 119
heat pumps 194, 199
hegemony 218
HFCS *see* high-fructose corn syrup
high fructose corn syrup 23, 36, 38
high-income nations 153
high modernism agriculture 13–16
high-yielding varieties (HYVs) 21, 23

Index   251

hogs *see* pork production
Hoover Dam 111
Hornborg, Alf 2, 4
human diversity 164
human environmental justice 2
human health factors, from livestock production 61
hydroelectric dams 108–116
hydroelectric power 187–188; Canada 115
hydrogen 191–192, 196–197
HYVs *see* high-yielding varieties

imperialism 22
incomes, guaranteed 151
India: farmers 63; nuclear energy 118; Untouchables 110
Indian corn 40
Indigenous Environmental Network 219
Indigenous peoples, Canada 99–102; *see also* First Nations, Canada
industrial agriculture, critique of 16–20
industrial farming 63; environmental degradation 70
Industrial Food Chain, versus peasant and small-farmer food webs 65–68
industry, renewable energy 195–196
Information Age 68
insecticides 19
insect pollinators 67
intergenerational buck-passing 208–210
intermodal freight 194
internal combustion vehicles 145
International Panel on Climate Change (IPCC) 224
international trade policies 5
internet 227
involuntary resettlement, dam constructions 108–110, 113–114
IPCC *see* International Panel on Climate Change
Iraq 94
Irish Potato Famine 16
irrigation 22
irrigation dams 108–116

Jackson, Wes 178–179, 223
Jacobson, Mark 182–183, 199–200
James Bay Project 115
John Deere 36
Johnston, Barbara Rose 4

Kariba Dam 109
Kellogg's 24
"King Corn" 48
Kissinger, Henry 23
Klein, Naomi 147
Klein, Seth 146, 159
Koch brothers 1
Kyoto Protocol (1993) 157

labor unions 80–81; United Mine Workers Union 85
land grabs 26–27
Land Institute 179
large-scale ranching 56
Latour, Bruno 216–219
La Via Campesina (LVC) 68–72, 74, 163, 165, 215
legislation, impact on future generations 212
The Leviathan 156–157
lithium 201
lithium-ion batteries 196–197
livestock operations 27
livestock production 49, 60–61; cattle ranching, Brazil 55–57; pork 57–60; rise of beef industry 49–55
lobbying governments 213
local ecosystems 17
localism 179
localization 10, 18, 222, 225
local resilience 167–168
Lovins, Amory 182–183
Lubicon Lake Cree 97
LVC *see* La Via Campesina

machine logic 18
machinery manufacturers 28
Mackenzie-Athabasca River drainage system 95
mad cow disease 70
maize 40, 66; *see also* corn
managed paddock grazing 174
markets 150
Marshall Islanders 1
Matewan Massacre 85
Mayan indigenous peoples 165
McDonalds 36, 46, 53–54
McKibben, Bill 223, 225
meat packing industry 53–54
mechanization 15
Megamachine 81

methane 30, 108, 111; from coal mines 81; decomposing biomass 173
Métis 102, 132
Mexico 24; corn 39–43
micro hydro power 187–188
Mikisew Cree First Nation 100
milpa ("three sisters" system) 164–165
minerals 99, 170, 202–203, 228; degrowth 155–156
mining coal 81; *see also* uranium mining
mission-oriented capitalism 160–161
mixed crops 226
modular production 165
monocrop systems 16, 176, 226
monocultural food systems 17, 68
monoculture 19
monopoly capitalism 28
morality regarding climate change 208–209
moral storms, how to respond 209
Mosaic 36
mountain-top removal coal mining (West Virginia) 83–90
movements, environmental and social 7, 8, 124, 125, 129, 132, 133, 214–221
mycorrhizal fungi 171
mycorrhizal symbiosis 20

NAFTA *see* North American Free Trade Agreement
national security 22
native prairies 178–179
natural gas, resulting from decaying vegetation 111
natural resources, degrowth 155–156
natural systems agriculture 175–180
Navajo reservation 102
negative-feedback loops, markets 150
negawatts 182
neo-classical economics 229
neoliberalism 229
nested localization, renewable energy 198–200
Netherlands, ecological footprints 226
net-zero carbon emissions 141–142, 144, 182; *see also* carbon dioxide emissions
New Deal programs 34, 223
New Green New Deals, cap and adapt 223
Nigeria, petroleum 94
Nikiforuk, Andrew 94, 202

Nile River 112
NIMBY (not-in-my-backyard) 129
nitrogen, livestock production 61
nitrogen fertilizer 23; corn 34
North American Free Trade Agreement 69; corn 39–43
not-in-my-backyard (NIMBY) 129
Nova Scotia, Sydney Tar Ponds 104
nuclear arms race 117
nuclear bomb testing 1
nuclear energy 117, 134–137; arguments for uranium mining 125–127; Coalition for a Clean Green Saskatchewan (CCGS) 127, 129–130; decommissioning nuclear reactors 135–137; problem of nuclear waste 130–131; release of Perrins 128–129; search for a repository in Canada 131–134; uranium mining 119–124
Nuclear Renaissance 123
nuclear waste 130–131, 183
Nuclear Wate Management Organization (NWMO) 128, 131–132
nuclear weapons programs 118
nutrition, impact of dams 112
NWMO *see* Nuclear Wate Management Organization

ocean power 188–189
oil *see* petroleum
Ontario Hydro 120
OPEC *see* Organization of Petroleum Exporting Countries
opposition: to tar sands 105–106; to uranium mining 124–125
Orano 121
Organization of Petroleum Exporting Countries (OPEC) 92–93
Orkney Archipelago, hydroelectric power 189
Oscar Meyer 53
overconsumption, Industrial Food Chain 68
overeating 180
over-fishing 61
overgrazing 174
ownership, shifting to usership 154–155

Pakistan, nuclear energy 118
Paris Agreement (2015) 157

pastureland, managed paddock grazing 174
pasturing ruminant animals 174
path dependencies 30–31, 135
Peace River 96, 115
peasants 64; La Via Campesina (LVC) 68–72
peasant webs 65–68
perennial grains 175–180
perennialism 179
permaculture 180
personal incomes guaranteed 151
PespiCo 24
pesticides 19; genetically modified organisms (GMOs) 67
petroleum 91–92, 214; petrostates 92–95; tar sands (Canada) 94–104
petrolization 92
petrostates 92–95
phase transitions 30–31
photovoltaic effect 185
Pinehouse First Nation (Canada) 133
pipelines 102–103; destroying 216
planned obsolescence 154
plant breeding 16, 48, 179
plant diversity 15–16
plants 169–170
plutonium 131
policies for implementing, Green New Deals 147–149
political ecology 216–217
political ecology and class 217
political economy 3–4, 6
political governance 157
political motives for Green Revolution 22–25
pollution, tar sands (Canada) 96
polycultural systems, where found 167–168
population 142
pork production 57–60
Post-development movements 152
Post-Development Pluriverse 158
power differences 4
Precambrian Shield 123
private financial investment sector 160
problem of nuclear waste 183
processed food 38
public advocacy 7
public will 219

pumped hydro 197
pyrolysis process 172

quality control, agriculture 25–26

railroads 80
Rainforest Action Network 8
ranching, cattle ranching (Brazil) 55–57
rationing 229
reclaiming the commons 228
redesigning the grid 198–200
refrigerants 195
regenerative/restorative agriculture 169–175
regulatory policies of governments 144
religious supporters of climate action 217
relocations, due to dam constructions 109–110
renewable energy 143, 182–183; bioenergy 190–191; for buildings 194–195; degrowth 156; electric vehicles 192–194; geothermal energy 189–190; hydroelectric power 187–188; hydrogen 191–192; in industry 195–196; ocean power 188–189; redesigning the grid 198–200; solar energy 184–185, 203; storing 196–197; stranded assets 201; wind power 186–187
renewable transitions in electricity production 184
resiliency 226
response to global warming and civilization collapse 225–229
Revenue Canada 121
rhizosphere 169–170, 177
Rickover Admiral Hyman Admiral 118
Rifkin, Jeremy 147
rights of individuals 208
right to repair laws 154
"Ring of Fire" 190
Rockefeller, Nelson 22–23
Rockefeller Foundation 22–23
Round-Up 44, 73
ruminant animals 174
run-of-the-river hydro 187–188
Russia, nuclear energy 117

sacrifice zone 84
salinization 22

Saskatchewan: agriculture 13; Coalition for a Clean Green Saskatchewan (CCGS) 129–130; Gunnar Uranium Mine 104; search for nuclear repository 131–134; uranium hearings 125; uranium mining 120–124, 128–129
Saskatchewan Environmental Society 124
Saskatchewan Party 122
Saudia Arabia 93
scaling down destructive industries 155
schistosomiasis 112
Scotland: Dounreay nuclear facility 135; wave power 189
Scott, James C. 14–15
sea ferries with renewable energy 193
second Green Revolution 70
self-determination 71
self-sufficiency 226
shale fracking 5
Shell 91
SIA *see* social impact assessment
Sierra Club 8, 105, 124
silent permission 220
silt build-up 111
Site C along the Peace River 115
small-farmer food webs 65–68
small-scale farmers 166–167, 181
small-scale hydro 187
small-scale political units 226
smart windows 195
Smil, Vaclav 200, 201
Smithfield 49
SNC-Lavalin 120
social impact: of dam construction 113–114; from livestock production 61–62
social impact assessment (SIA) 98
social impact studies 7
soda drink industry 37–38
soil quality 169–175
solar energy 184–185, 203
solar photovoltaics (PV) 185
solutions to climate change 161–162; agroecology 163–169; buck-passing 207–208; degrowth 149–156; food waste 180; Green Growth 142–145; Green New Deal 145–149; natural systems agriculture 175–180; perennial grains 175–180; permaculture 180; regenerative or restorative agriculture 169–175; responsibility for 156–158; small-scale farmers 181; *see also* renewable energy

Soviet Union, nuclear energy 118
soy 43–47; oils 23
Spain, wind power 186
Standard Oil 23
steady-state economies 222
steam engines 80
storing renewable energy 196–197
stranded assets 201
strikes, labor unions 80–81
subsidiarity 225
sufficiency 222, 226
sun 184
Suncor Energy 214
Sunrise Movement 211
"supermarket tomato" 16
sustainability 153, 164–166
Swift 53
Sydney Tar Ponds 104
Syngenta 47, 64

tar sands (Canada) 94–104; sterilization of 105
tax credits 147–149
taxes, degrowth 151
tax rebates 147
"Taylorism" principles 15
temperature rise 224
terra preta (black earth) 172
thermochromic windows 195
The Third Act 218–219
thorium 135
Three Gorges Hydro Complex 189
Three-Mile nuclear plant 124
tidal energy 188–189
tools for conviviality 227–228
tortilla prices 42
trains, electric trains 193
trams, electric vehicles 193
transgenic crops 73
transition discourses 152, 200–201
transition towns 227
transportation 199; electric vehicles 192–194
Treaty 8 97, 99
trucking 194
Tyson Foods 49

UDP *see* Uranium Development Partnership
United Kingdom, plutonium 131
United Mine Workers Union 85
United States: agriculture 23–24; beef industry 51–55; buildings 195; control

of food 63–64; corn 39–43; dams 114; Green New Deal 161; imperialism 22; petroleum 94; renewable energy 183; soy 44–47; Sunrise Movement 211; wind power 186
Untouchables, India 110
Uranium Development Partnership (UDP) 122
uranium mining 102; arguments for and against 125–127; in Canada 119–120; future of 128–129; moments of resistance 124–125; in Saskatchewan 120–125
Uranium oxide (U3O8) 121
urban agriculture 66
urbanization 180
U.S. Atomic Energy Commission (USAEC) 117
US nuclear bomb testing 1
USAEC *see* U.S. Atomic Energy Commission
usership (rather than ownership)154–155

value chains, agriculture 25–26
virtual extinction of American buffalo 49–55

Walk for 7000 Generations 133
warm season grasses 179

war on bugs 16, 166
wasted food *see* food waste
wastelanding 102
water 5, 22, 29; high-yielding varieties (HYVs) 21; pork production 60; tar sands (Canada) 96
water reservoirs 111
wave energy converters 188
wave power 188–189
wealth, shifting of 153
West Virginia, mountain-top removal coal mining 83–89
wet mills 36
wheatgrass 179
wind farms 199
wind power 116, 186–187
wind tower blades 186
wind turbines, minerals 202
Wood Buffalo National Park 100
World Bank 110
World Economic System 4
World Trade Organization 24–25, 69
World War II 146, 159
WWS world (wind, water, and solar) 182, 191, 199–200, 203

zero-tillage 173

For Product Safety Concerns and Information please contact our EU representative GPSR@taylorandfrancis.com Taylor & Francis Verlag GmbH, Kaufingerstraße 24, 80331 München, Germany

Printed and bound by CPI Group (UK) Ltd, Croydon, CR0 4YY
08/06/2025
01897008-0004